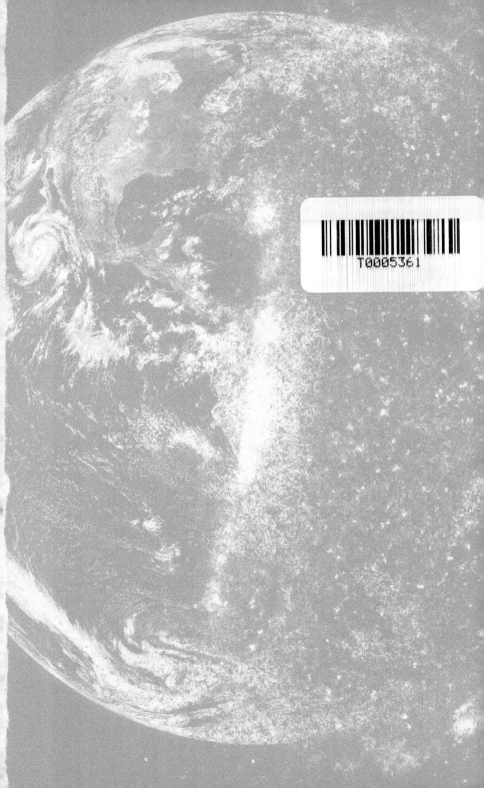

Advance Praise for
Climate Cult

"Brian Sussman does an excellent job of showing carbon dioxide is not a major factor in the earth's temperature, therefore, why demonize fossil fuels? This raises an important question; what is driving the gigantic international support that claims humanity is forcing climate change? The answer can be found in the opening statement of the first international conference on global warming in Rio De Janeiro at the first Earth Summit in 1992. The late Maurice Strong, Chairman of the meeting, sounded a false alarm that climate change 'could lead to the end of our civilization.'

"Nearly two decades later, Ottmar Edenhofer, a co-chair of the United Nations Intergovernmental Panel on Climate Change, explained the real intentions of this movement, stating, 'But one must explicitly say: we de facto redistribute the world's wealth due to climate politics… One has to free oneself from the illusion that international climate politics is environmental politics. This has almost nothing to do any more with environmental politics…'

"There it is, in plain sight, and it has nothing to do with carbon dioxide, which they've deemed 'evil.' CO_2 is being used as an instrument to destroy our economy and force the US into a United Nations' One World Marxist type government. Indeed, the initial purpose of that inaugural Earth Summit was to assemble a 'parliament of the planet' in the name of 'Only One Earth.'

"Sussman provides us with the best understanding of this truth than any of the other authors of over thirty books I have read by those skeptical of anthropogenic climate change. He traces these environmental contentions all the way back to Karl Marx and Vladimir Lenin. If you are interested in exploring the answer to the question I posed at the beginning of my comments, then *Climate Cult* is a must-read."
—**Dr. Neil Frank,** Former Director, National Hurricane Center

"Brian Sussman's *Climate Cult* is just the anecdote for those who need deprogramming from the irrational fear of climate change. This book goes deep into how the high priests of climate change are seeking to use unfounded climate fear as a means to reorder society and centrally plan your life. Sussman is uniquely qualified to expose how normal climate and weather are used as muscle against you. Before you drink the Kool-Aid, please take a big gulp of Sussman's *Climate Cult*."

—**Marc Morano,** Publisher, Climate Depot

"Brian Sussman was a meteorologist and rising media personality who was 'in the room' when his industry leapt to promote what was, at the time, *global warming* catastrophism. Declining to go along with what he convincingly describes as the 'climate cult,' Sussman took professional risk while showing personal courage. He stood up and spoke out. The debate—a debate that rages, as affirmed by endless public polling despite widespread, ahem, *denial* of that fact—is better for it. Now, Sussman dissects the cult-like aspects of the 'climate' movement, which is also very big business, an industry that inherently demands we surrender our freedoms and prosperity. We took too much for granted, and too much on faith, as this industry grew. Brian Sussman does his part to make sure we are unlikely to be fooled again.

—**Christopher Horner,** Author of *The Politically Incorrect Guide to Global Warming (and Environmentalism)*

"The greatest threat to freedom in the world is the Climate Cult. This movement of religious zealots that worship the created, rather than the creator, has met their match in this brilliant book by Brian Sussman. Page by page and word by word, Brian shows how this movement is dangerous to civilization. The civilized world depends on energy and these cranks have made it their mission to make energy unaffordable to all but the very rich."

—**Floyd Brown,** Author of *Counterpunch,* Founder of The Western Journal

"I have been screaming about the climate cult for years. Finally, a book that exposes the red roots of the 'green' agenda."

—**Chris Salcedo,** Newsmax

CLIMATE CULT

CLIMATE CULT

CULT

EXPOSING AND DEFEATING
THEIR WAR ON
LIFE, LIBERTY, AND PROPERTY

BRIAN SUSSMAN

Post Hill
PRESS

A POST HILL PRESS BOOK
ISBN: 979-8-88845-544-9
ISBN (eBook): 979-8-88845-545-6

Climate Cult:
Exposing and Defeating Their War on Life, Liberty, and Property
© 2024 by Brian Sussman
All Rights Reserved

Cover design by Cody Corcoran

Post Hill Press
New York • Nashville
posthillpress.com

Published in the United States of America
1 2 3 4 5 6 7 8 9 10

∞

Dedicated to the legacy of those who pledged
their lives, fortunes, and sacred honor.

∞

TABLE OF CONTENTS

INTRODUCTION

1996. WEATHER DESK, TV CHANNEL 5's *Eyewitness News*, San Francisco.

"Hello. Brian Sussman here."

On the other end of the line was a newspaper writer in Florida, congratulating me on some awards I had recently received, while throwing in some niceties about my tenure on an American Meteorological Society educational committee as well. Schmoozing is the technical term I'd always used for that type of over-the-top flattery, especially from a stranger, and immediately flags of suspicion snapped over the phone line. But by the time my thoughts and words could coalesce into a sentence, the caller demanded, "As a television meteorologist, why don't you believe in global warming?"

Without a breath, he began ripping into me for endorsing the Leipzig Declaration, a public document skeptical of human-caused, or anthropogenic, global warming, drafted in opposition to the United Nations' 1995 Climate Treaty. The heart of the declaration states:

> We believe that dire predictions of a future warming have not been validated by the historic climate record, which appears to be dominated by natural fluctuations, showing both warming and cooling.

"I agree with the declaration," I calmly said into the phone.

"You don't believe humans are influencing the climate?"

"I don't think we can be sure."

"So, you don't believe there's global warming."

"The jury's still out."

The reporter relaunched, claiming there was "a consensus among *real* scientists" and that I should be disqualified from signing on to such "propaganda" because I was a "*television weatherman.*"

I politely ended the call, wondering, *What kind of crappy excuse for journalism was that?*

For the record, Leipzig was signed by some twenty television meteorologists as well as eighty PhD scientists—all significant players in their respective fields of atmospheric science, geoscience, oceanography, and physics. One of the experts was even instrumental in the development of the first weather satellite, another was a research physical scientist at the National Climatic Data Center, and a third ran the University of Hawaii sea level program.

Other disturbing encounters soon followed that eventually caused me to sense a religious fervor among many who presume the theory of anthropogenic global warming is an unalterable fact.

For several years while at *Eyewitness News*, I had also been the regular fill-in weatherman for the nationally broadcast *CBS This Morning* show in New York City. The iconic anchorman for the *CBS Evening News*, Dan Rather, was in San Francisco doing his newscast from our San Francisco station one week. We bumped into one another in the studio, and after a pleasant back-and-forth, he pitched me on an exciting plan.

"We need a West Coast weather correspondent on the *Evening News*; I think you'd be the perfect guy."

Floored, I replied, "Wow. What an honor, Dan."

"And you know, I think you would be the perfect guy to talk to our audience about global warming," he continued.

"That would be great—of course, the verdict's still out on global warming, and we'll need to sort the discussion out. But, yes, I'd love to be available."

Dan really didn't respond, other than to quickly excuse himself and head to the dressing room. And obviously, I never made it to the *CBS Evening News*.

A third encounter followed the next year after the West Coast had just experienced an incredibly active winter due to a combination of an El Niño warming of the Pacific Ocean and a common weather feature often referred to as the "Pineapple Express," a plume of moisture that follows the jet stream from the tropics towards the poles, dumping lots of rain and snow in the process,[1] which prompted Vice President Al Gore to call for an El Niño summit in Santa Barbara, where he could be among the first to try link El Niño with global warming.

The TV station tapped me to anchor live coverage of the vice president's visit.

---

1   The Pineapple Express weather pattern is a perfectly normal meteorological feature. It has various nicknames in different parts of the world, but technically it is a plume of moisture being carried from the tropics to the higher latitudes by a jet stream. These patterns are an important feature of the planet's seasonal weather and overall climate, as they transport some 90 percent of the moisture from the tropics to other parts of the world. In the last couple of years, this perfectly normal and necessary weather pattern has taken on a new moniker, one with near biblical implications: *atmospheric river*, a term Al Gore has taken a great liking to.

I had no problem with that, except for one important item. I told my boss I was not going to attribute El Niño to global warming.

"It's the biggest weather story in years," he insisted.

"There's no scientific evidence for it."

He rolled his eyes. "You're in denial."

He then leaned towards me and, lowering his voice, said, "Brian, you know what your problem is? You don't know which facts to leave out."

My blood pressure boiled, but I held my tongue. Given the fact that this guy had been the president of his college's Students for a Democratic Society—a brazen socialist organization—his arrogant comment didn't come as a total surprise. I just couldn't believe he actually said it out loud.

The newsroom did send me to Santa Barbara to present a primer on El Niño, along with my regular weather cast, but not to cover Al Gore or the conference.

In 2001, with the addition of our fourth child, I worked out a deal with CBS to exit my contract midstream so I could be home in the evenings with our growing family. The TV news biz was rapidly changing, as were the attitudes of those who believed in anthropogenic global warming. It was an easy decision to make.

Serendipity then led to an offer to host a talk show on an immensely popular radio station in San Francisco. I accepted the opportunity but quickly endured even more blowback from global warming believers who were bizarrely enraged that the affable, former weather guy was now an official "denier," spouting lies and misinformation over the radio airwaves.

On-air debates with callers and nasty emails from listeners got my dander up enough to double down on my research and dig into the debate, which resulted in my book *Climategate: A Veteran Meteorologist Exposes the Global Warming Scam* in 2010. That exposé scientifically debunked all the popular global warming claims while revealing the detailed socialist agenda behind them and their attempts to utilize climate as a weapon to attack capitalism, free enterprise, and our American way of life. A sequel followed in 2012, *Eco-Tyranny: How the Left's Green Agenda Will Dismantle America*, that further exposed the freedom-robbing intentions of the entire environmental movement at large.

By then, some friends and many more acquaintances shunned me straight-out. Most of my former TV colleagues acted as if I had a plague. It was troubling yet curious. *Their* personal views on climate didn't cause me to dislike *them*—why, then, I wondered, did *my* views on the topic foment such high animosity towards me? But my naiveté dissipated quickly with blatant forms of censorship infiltrating my social media accounts from companies like Facebook and Twitter. Stunned at first, I soon had my first real "cognitive dissonance" moment when I realized my intuition was, indeed, accurate: I was dealing with a cult, a bona fide religion with a central tenant and high priests who will stop at nothing to proselytize their doctrine on all humanity by any means and at all costs.

Around the same time, I was invited to present a guest lecture on global warming before several hundred students at the University of California, Berkeley. It wasn't long into my talk I was listing facts that contradicted the students' educational experience. Their agitation was clearly obvious, and, thankfully, the professor who had invited me calmed everyone down before things got out of hand.

After the presentation, I informally answered questions with some of the students in front of the stage. Our exchange was revealing. Most of them were dismayed—even shocked—that they had never heard my compelling side of the climate change argument before. They felt their educators had ripped them off on this subject. It was if my talk had awakened them from an intense brainwashing.

A second speaking engagement sponsored by the Republican Club at San Francisco State University was the deal-clincher for me in terms of my understanding what exactly I was dealing with in the phony climate change debate.

As I walked onto the stage that day, I was greeted by a large auditorium filled with the backs of countless students who sadly thought they were making a significant point. The Republican Club students filled a couple of front rows and did welcome me warmly.

Okay. So the others froze me out. What, though, was the point the about-face crowd was making? In my view, they weren't open to learning, to examining the facts on both sides of the issue. Heck, they weren't even open to listening before disagreeing with me.

Long story short, aside from the brave young Republicans, it was a miserable experience. I ended up having to be escorted off campus by law enforcement.

By that time, I was absolutely, positively sure—*the climate change crowd is a cult.*

A WORD TO THE COMPASSIONATE

It is hard for many of us to not remember a time before 9/11 or to have no concept of a world where you couldn't "Google it." However, for Generation Z, those born from 1995 onwards, the only memories of 9/11 come from documentaries that they may have seen—and a world without Google? Incomprehensible.

My generation—the baby boomers—grew up under parents who were products of the Great Depression and World War II. Many were first-generation Americans. The adults in our lives were dubbed, "the silent generation," as they stuffed their emotions while striving for the American Dream. A large percentage had been raised in poverty, too many had experienced the horrors of war, but they all saw the United States as a champion of liberty and were unabashedly proud of it. Their perspective produced hard work, large families, white picket fences, and suburbia. While many boomers embraced their parents' values, others adopted a significantly more nuanced view of the USA.

The offspring of my generation, the millennials, found themselves coming to age in the new era of opinionated TV talking heads, fiery talk radio hosts, and the polarizing presidencies of Bill Clinton, George W. Bush, and Barack Obama. Given that social and alternative media was only in its infancy, millennials received whatever their parents were choosing to consume as well as what their favorite school teachers were preaching. However, some, particularly the younger portion of that group, became immersed in what the older generations feared: technology. It required new ways of thinking, seemed to breed selfishness and laziness, flirted with immorality, and divulged far too much personal information. But for Gen Z, utilizing tech to participate in a variety of groundbreaking platforms quickly became an exciting lifestyle, allowing them to experience the world as their virtual oyster. This generation feels connected to the entire planet, causing them to experience unprecedented empathy for others like no other age group in history. Gen Z, and many millennials, are a compassionate people, truly desiring to help others, love others, and save the world from polluting the environment and harming the climate.

However, their very real emotions have been played upon by nefarious actors in high places who are looking for what are known in political jargon as "useful idiots." In many ways, I'm writing this book for those much younger than I. My earnest desire is to present a facts-based perspective that will hopefully cause many to step back and honestly reexamine their strongly held convictions that the climate is out of control, and drastic measures including new laws, mandates, regulations, and peer pressure must be instituted to change the course and reset society. To those in that camp, I urge you, please be willing to hear me out. I'm not trying to offend anyone. My research is well-sourced, and I dearly desire to enlighten you with facts you've perhaps never been aware of.

THEOLOGY OF ECOLOGY

That said, I'd like you to think about this: the premise of climate change has become similar to a system of ecclesiastic belief, based on pseudo-facts and science, while manifesting traditional religious elements of guilt, revelation, repentance, devotion, and duty. It's as if there has been a fall from original grace, and the once perfect Mother Earth has been corrupted by greedy carbon sinners. Leaders of the cult preach dreams of a new world, a pristine environment, a calmed climate, a manageable population, and universal salvation through sustainable development, which, in turn, will expunge the world of social injustice and inequity.

This cult has its own prophets and evangelists dedicated to converting the masses through the cataclysmic messaging of fear. Its teachers and influencers catechize the faithful, shaping and influencing their behavior. Climate doctrine holds that the environment has been despoiled by the by-products of luxurious living. Guilt is repressed through acknowledgment of such allurement and the subsequent pursuit of a frugal existence and a reduced carbon footprint; these are all paths of self-righteousness and outward virtue.

Businesses can find absolution as well by planting trees to offset their grubby footprint. Governments atone by passing laws to ensure that more energy comes from solar and wind power. The wealthy can make up for that large second home on the lake by dropping $135,000 on a loaded Tesla X—while ignoring the toxicity of its 7,104 lithium batteries[2] required to power it.

It is a theology of not just ecology but sociology and philosophy as well. Beginning in kindergarten and continuing into college, a daily, interdisciplinary, educational diet of green tenets has been infused into a collective psyche. The cult's infernal enemies are capitalism and, notably, the American ethos of liberty. Unified, never-ending media communications incessantly advance the indoctrination. Those who oppose the tenants of

---

2   Denis Muroki, "How Many Batteries Are There in Each Tesla Model," *History-Computer*, April 11, 2023, https://history-computer.com/how-many-batteries-in-each-tesla-model/.

this faith are shunned and censored as apostates, their fates determined by their own choices.

Even traditional religious leaders have been deceived and drawn in, adding additional plausibility to this sect. "It is our sister, Mother Earth, who cries out," declared Pope Francis. "Prey to our consumerist excesses, she weeps and implores us to put an end to our abuses and to her destruction."[3]

Describing climate change as a cult does not necessarily imply that anthropogenic climate change or global warming does not exist. Indeed, some may insist the evidence for an altered atmosphere is overwhelming, while others, as you will discover, are cautiously skeptical or even staunch non-believers. However, no matter its empirical basis, climate change is progressively taking the form of spiritual orthodoxy, not only fulfilling some of the individual needs associated with such a belief system but dangerously enabling political leaders and policy makers to move forward with an exhaustive agenda requiring more laws, statutes, rulings, and entire bureaucracies that directly chip away at the inalienable rights established in the United States' founding documents.

What the climate change agenda lacks in science it makes up with highly dramatic rhetoric and claims, premeditated to frighten energy sinners to repentance. For example, Prince Charles told an audience of business executives and influential environmentalists, "In failing the Earth, we are failing humanity." He warned the next generation will face a "living hell" without government intervention. "Our consumerist society comes at an

---

3   "Pope Francis urges world leaders to act on extreme weather," *Aljazeera*, July 21, 2022, https://www.aljazeera.com/news/2022/7/21/pope-francis-calls-on-world-leaders-to-act-on-extreme-weather.

enormous cost to the Earth and we must face up to the fact that the Earth cannot afford to support it."[4]

Charles then uttered his prediction: we have just "96 months"—eight years—to avert "irretrievable climate and ecosystem collapse, and all that goes with it."

The man who is now king made his prediction in 2009. However, in 2015, Charles extended his deadline for the planet's demise by giving it another thirty-three years. King Charles is not the only eco-prophet to amend a doomsday prophecy or climate omen. Later in this book, you'll discover a myriad of them.

If, by chance, a climate Armageddon cannot be averted, the cult's more pedigreed leaders tell us they possess the knowledge to effectively reengineer the complex ways of the atmosphere. But, so far, they say they would *never* do something as controversial as that.

Meanwhile, the run-of-the-mill congregants seated in the back rows of the climate temple are distracted by their TikTok feed and lulled into a sense of contentment, completely oblivious to a well-planned authoritarian ideology that has miserably failed countless times throughout human history.

The climate agenda is as much about counterfeit social justice and social equity as it is anything else. While I understand there is injustice and inequality in this world—including in the United States of America—my comments on these important topics are rooted in the fact that the foundational concepts of justice and equity have been hijacked and redefined by progressive

---

4   Robert Verkaik, "Just 96 months to save the world, says Prince Charles," *Independent*, July 9, 2009, https://www.independent.co.uk/climate-change/news/just-96-months-to-save-world-says-prince-charles-1738049.html.

philosophers, nongovernmental organizations, and governments themselves—and the rank and file have bought in.

In the chapters to come, I will carefully lay out the complete canon of the climate cult, from its crafty genesis to its planned revelation of paradise on Earth. To defeat this cunning scheme, which has stretched its tentacles deep into the fabric of America, we must fully understand its roots and bring to light all of its vengeful programs designed to usurp life, liberty, and the happiness realized in the ownership of personal property, both physical and intellectual. Then, we can properly take our stand, distinctly make our voices heard, cause this enemy of the people to know we're on to their designs, and right the ship.

CHAPTER ONE

GENESIS

TO FULLY GRASP THE THINKING of the climate cult, the ideas of two diametrically opposed philosophers must be understood: Karl Marx, architect of socialism and communism, and John Locke, designer of classic liberalism (not to be confused with classic liberalism's third cousin, *progressive* liberalism). Marx's theory in a nutshell states, "Communism is the riddle of history solved, and it knows itself to be this solution."[1] Locke's enduring motto declares, "Life, Liberty, and the pursuit of Happiness."

Karl Marx was born in Germany in 1818 and died in England in 1883, a relatively short sojourn but certainly an ample amount of time to conceive an intricate philosophy he originally dubbed, "organized collectivism."

In his teens, Marx was drawn to political liberalism. His rebellion was further confirmed while studying at the University of Berlin where he was strongly influenced by radical thinker Georg Hegel. Hegel believed Christianity had a negative effect on society akin to an illness and that a new type of religion built on scientific reason was needed. Marx was so taken by Hegel that

---

1   Karl Marx, *Private Property and Communism* (1844), https://www.marxists.org/archive/marx/works/1844/manuscripts/comm.htm.

he joined a group on campus known as the Young Hegelians. Their initial goal was straightforward: liquidate Christianity.

Hegel lectured that everything in the universe could be explained through his system of rational thinking known as the "dialectical process." Originally conceived by Greek philosophers, Hegel's version held that contradictions in nature do not harm one another but instead lead to a higher level of development, particularly personal intellectual development—hence, no need for religion.

Marx would eventually take Hegel's dialectic theory and refine it for philosophic purposes. Regarding religion, Marx would eventually speak of man's imaginary or "fantastic reality of heaven, where he sought a superman...."[2] The demand for "real happiness," Marx taught, was found in "the abolition of religion."[3]

In 1841, Marx received a doctorate in philosophy, having written his thesis on Epicurus, the ancient Greek theorist and quasi-atheist who taught that although the gods exist, they are not involved in human affairs; upon death, life ends, and there is no afterlife—the physical world is all there is and all there will ever be. Epicurus also developed a thesis on matter that contends that the fundamental constituents of the earth's system are invisible, indivisible bits of matter known as "atoms."

By Marx's time, Epicurus's theory of matter had become known as "materialism," and Marx was, indeed, a true believer.

2   Karl Marx, *A Contribution to the Critique of Hegel's Philosophy of Right*, "Introduction," (1843), https://www.marxists.org/archive/marx/works/1843/critique-hpr/intro.htm.
3   Ibid.

MARX'S LAW

In 1842, through correspondence, Marx met German philosopher, political theorist, historian, journalist, revolutionary socialist, and businessman Friedrich Engels. After finally meeting in person in Paris in 1844, the two became close friends and collaborators, developing a holistic ideology that would eventually change the world for the worse, bringing upon it inconceivable waves of human tragedy. Holding parallel views on materialism, dialectics, and the abolition of God, the two were convinced science was the ultimate path to preeminence. "If science can get to know all there is to know about matter," they proclaimed, "we will then know all there is to know about everything."[4]

For Marx and Engels, matter—atoms, molecules, and the otherwise unseen—was the alpha and omega of reality. Matter provided the complete explanation for plants, animals, man, intelligence, planets, and solar systems. Time was the magic wand that allowed all matter to coalesce, creating the universe in which humans find themselves.

To codify their doctrine, Marx and Engels prescribed their three laws of matter: the law of opposites, the law of negation, and the law of transformation. Like Moses coming down from Mt. Sinai with the Ten Commandments etched into stone, Marx and Engels's "laws" provided the rationale for communist doctrine. Those same "laws" also inspire the contemporary environmental movement's leading influencers and planners, specifically when it comes to the climate agenda.

---

4   Friedrich Engels, *Ludwig Feuerbach and the End of Classical German Philosophy* (1886).

- The law of opposites is an extension of Georg Hegel's original work, supposedly illustrating how everything in existence is a combination of dialectics working in unity. For example, electricity is characterized by a positive and negative charge. Likewise, atoms include protons and electrons, which are contradictory forces working in unity. Even the human race is composed of opposite qualities: altruism and selfishness, courage and cowardice, humility and pride, masculinity and femininity. To function properly, Marx and Engels believed these opposite forces must be kept in balance; if not, discord will be the result. Thus, the law of opposites demands *humans* be kept in check because, as the most advanced creatures, they are capable of wreaking the most havoc, hence, the need for a tightly regulated, often heavy-handed, system of government.

This first law concludes authoritarianism as essential to effective, masterful governance.

- The law of negation provides a key pillar for environmentalists. It notes that all species possess an inherent tendency to proliferate. However, while Marx and Engels believed that nonhuman species bear automatic mechanisms for manageable growth, the extended family of *homo sapiens* is incapable of similar self-regulation. Thus, negation casts humanity as a largely ignorant, ever-consuming population bomb placing the entire planet at risk.

As a result, negation insists systems must be properly arranged to maintain sustainability, including mechanisms to assure human population control.

A variant of the law of negation is the eco-buzzword we are constantly bludgeoned with today: *sustainable development*, or a call for government policies to force changes in human behavior and lifestyles under penalty of law. More on that later.

The third Marx-Engels's axiom is the most arrogant: the law of transformation. It states that continuous quantitative development by a particular species often results in a "leap" within nature whereby a completely new form or entity is produced. Transformation was bolstered by the findings of Marx and Engels's contemporary, evolutionist Charles Darwin. Darwin's theory resonated with the two, sealing their convictions that such "leaps" not only allow for the origin of new species but could produce a leap *within* a species, particularly within *homo sapiens*. This could enable some humans to advance to new levels of reality. The law of transformation contends there is an elite status within the human race, and those born into evolution's aristocracy possess a duty to dictate how the underdeveloped, or lesser-minded, shall live.

Taken to an extreme, transformation may also determine who shall die.

With these new revelations, Marx and Engels arrogantly boasted, "The last vestige of a Creator external to the world is obliterated."[5]

Here is my summary of the three laws of matter:

---

5   Friedrich Engels, *Anti-Dühring* (1878).

- Love, passion, value, and feelings are not composed of matter and are therefore imaginary. Belief in God is pure fantasy. Even consciousness is simply the result of material interactions within the human brain.

- The human race is naturally inclined toward overpopulation and incapable of peace and harmony without intervention from those born with a leap of intelligence above and beyond the masses. Those superior beings have a responsibility to collectively rule over those with less intellect, otherwise the lesser-minded will destroy the planet and kill one another.

- Thus, power must rest in the hands of a very few, assuring that the working class is kept content and compliant. Progressive laws, regulations, ethics, and ethos must be promoted and enforced.

Over the ensuing years, Marx never drifted from his materialistic assumptions and antagonist view of Christianity. Instead, he was able to neatly tuck those ingredients into his theory of organized collectivism, also known as socialism and communism (Marx used the terms interchangeably). He defended his philosophy as the establishment of a holistic framework from which society could be rebirthed and utopia established.

Today, such a patrician worldview resonates with those who consider themselves to be progressive and broad-minded, and highly resounds with those who have been trained at elite colleges and universities. Through advanced education and absorption of forward-thinking doctrines, this clique surmises they are masters of the universe, keepers of an amoral system wherein they have the power to define the rules, create the laws, determine the morality, and decree the rights of the people.

Hence, America's unique inalienable rights of life, liberty, and the pursuit of happiness are viewed as completely absurd, in large measure because an imaginary God cannot declare rights. Marxism demands that all so-called "rights" be issued by the powers of government and withdrawn by that same government if deemed necessary.

Historically, the philosophy of Marx has created some of the world's most offensive villains, while the philosophy of John Locke has inspired and empowered the world's greatest heroes.

NATURAL LAW, NATURAL RIGHTS

John Locke was known as "The Great" John Locke, and he is certainly one of the most important (albeit largely unknown) names in American history.

Locke lived over a century before Marx, from 1632 to 1704. He was an English philosopher, educator, government official, physician, and theologian who became very influential among the founders of the United States even though Locke himself never visited the American continent.

By all accounts, John Locke was an absolute gentleman. Raised in a village outside of London, he fondly described his mother as "pious" and said he was "in awe" of his father. He held to a Puritan Christian faith and was well-known for personal integrity, temperance, and friendship. A graduate of Oxford, he eventually became a professor there.

In the countless debates leading up to the drafting of America's Declaration of Independence and Constitution, Locke's scholarly wisdom was quoted regularly by our founding fathers. Physicist, inventor, and signer of the Declaration, Benjamin Franklin, said Locke was one of "the best English authors" for the study of "history, rhetoric, logic, moral and natural philosophy."

For the sake of this discussion, Locke's notable developments were that of natural law, natural rights, and their by-product, liberty. In examining the roots of the climate cult, we must properly define these terms, as they are not only inconsistent with Marxism and the climate agenda, but they are under vicious attack in America today.

The concept of natural law, or the law of nature, was well understood in Locke's time, but no one had yet applied it to the system of representative government that Locke was developing. Quite simply, natural law holds that our Creator has divinely placed within our hearts the knowledge of right and wrong, good and evil, truth and lies. It is a universal code of ethics succinctly expressed in the Bible's Ten Commandments. English jurist Edward Coke (1552–1634) efficiently summarized natural law:

> The Law of Nature is that which God at the time
> of creation of the nature of man infused into his
> heart, for his preservation and direction: and this is
> lex aeterna [the eternal law], the Moral Law, called
> also the Law of Nature. And by this Law, written
> with the finger of God in the heart of man, were
> the people of God a long time governed, before
> that Law was written by Moses, who was the first
> Reporter or Writer of Law in the world.[6]

---

6   Andrew Greenwall, "Pursuit of Happiness: Edward Coke and the Declaration of Independence," Lex Christianorum, May 17, 2011, http://lexchristianorum. blogspot.com/2011/05/pursuit-of-happiness-sir-edward-coke.html.

Natural law declares a system of morality without which there would be no bounds on human action. Locke believed these basic laws of nature should be protected from government interference and encouraged by the people.

In addition to natural law, Locke espoused natural rights. He identified men and women as the workmanship of God from whom they received three distinct rights: life, liberty, and the happiness derived from the ownership of property; this is where the Declaration of Independence's phrase "the pursuit of happiness" comes from.

These natural rights are inalienable (or *un*alienable—same thing). In other words, they have been given to humanity by their Creator and, therefore, should not be restricted. Locke insisted the only legitimate end of government was the defense of these natural rights.

Keep in mind how unique this was—and still is. Prior to the founding of the United States, all nations had been ruled by monarchies, oligarchies, dictators, or military leaders. Locke's theories implied a nation's laws be congruent with the moral construct of natural law, and that natural rights be protected by a constitution. A superior constitution, Locke argued, should allow for a limited government elected by the people, complete with a distinct separation of powers—no king or queen, no authoritarian leader.

However, Locke cautiously warned that legislative power must be "a power that has no other end but preservation, and therefore can never have a right to destroy, enslave, or designedly to impoverish the subjects."[7]

---

7   John Locke, *Second Treatise on Civil Government*, Chapter 11, Section 135 (1689).

Not to mention curb independent thought and individualism.

Immediately we distinguish the marked differences between Marx's organized collectivism and the system of governance contemplated by Locke. Whereas communism is pessimistic, untrusting of humankind at large, and stifling to independent thought and individualism, Locke's philosophy tends to be optimistic, believing in the best of mankind, while encouraging individual responsibility and achievement.

LIBERTY AND HAPPINESS

Locke, in his popular 1690 essay, "Concerning Human Understanding," wrote, "The necessity of pursuing happiness [is] the foundation of liberty."

Everyone pretends to be a champion of liberty, including progressive liberals, socialists, and communists who get quite offended when accused of being an opponent of freedom. However, once again, we must precisely define the terms. Both liberty and the pursuit of happiness do not equate to doing your own thing. The contentment spoken of by Locke stems from living within the bounds of natural law and finding joy in the natural rights, particularly the happiness associated with property ownership.

The real meaning of *the pursuit of happiness* has been lost in America. As I write this, two weeks ago I was speaking to a large pro-life gathering in California. My speeches are always lively, highlighted with both good humor and hard facts. The audience was a delightful group who put their money where their hearts are on the issue of abortion. In the course of my presentation I explained, "Our natural rights are life, liberty, and the pursuit of happiness. The latter right implies the pursuit of property ownership."

I was shocked by the reaction as I finished that sentence.

About a quarter of the audience laughed, as if they thought I was setting up a joke.

It's bothered me quite a bit, but I'm guessing those who laughed are victims of an education system that never properly taught the foundational concept of inalienable rights. I realize I was speaking in a very liberal city in California, but still, it was stunning and not in keeping with the spirit of what Locke and America's founders believed. Property doesn't just extend to physical possessions but also to intellectual values and the work of your own hands. One finds contentment when his hard work is compensated, and the money earned is used to purchase his own food, clothing, and the roof over his head—no matter how humble the home. When someone buys a bike or a used car, there is a sense of joy; the same can be said with the acquisition of all sorts of material possessions. People are thrilled when they purchase a house, farm, or commercial property. And of course, there is the excitement one has between his or her ears when it comes to entrepreneurship, personal goals, dreams, and planning for the future—these, too, are one's property. John Locke described it this way: "Every man has a 'property' in his own person."[8]

Liberty is the advancement and protection of the inalienable natural rights of life and property. In fact, Locke also made a tenacious argument that mankind has a sacred right to physically defend his property, giving rise to the Second Amendment's right

---

8   Ibid., Chapter 5, Section 27.

to bear arms.[9] There is no escaping the fact that America was founded on the principle that property is synonymous with liberty and security. Our founders envisioned the United States as a haven for happiness, believing that in a free-market economic system, void of government overreaching and central regulation, a new worker or immigrant could climb the class ladder in conjunction with their ingenuity and work efforts and reap the benefits associated with owning their own business, farm, bank account, home, and estate, not to mention holding fast to their personal philosophical and religious beliefs.

As Locke brilliantly stated, "God, who hath given the world to men in common, hath also given them reason to make use of it to the best advantage of life, and convenience. The earth, and all that is therein, is given to men for the support and comfort of their being."[10]

The right to pursuing the happiness associated with one's property is ideally to be conducted within the framework of natural law, specifically the Golden Rule's command: "In everything, therefore, treat people the same way you want them to treat you, for this is the Law and the Prophets."[11]

Locke and the founders were not espousing a system built upon selfish greed, though, given human nature, they knew that would be a risk. But they also felt if someone seeks abundant riches and obtains much in a lawful manner, that should be their own prerogative.

---

9   Ibid., Chapter 11, Section 136.
10   Ibid., Chapter 5, Section 26.
11   Jesus, speaking in Matthew 7:12 (New American Standard Version).

SELF-GOVERNANCE

Liberty also implies the right to self-governance. Again, a monarchy, oligarchy, or authoritarian ruler was never the plan for the USA. As the opening words of the Declaration boldly state, self-governance is "We the People," or, as President Abraham Lincoln articulated in 1863, "Government of the people, by the people, for the people."

Liberty is the citizen's guarantee of freedom from tyranny.

Let me reveal a great contrast, quoting directly from the Communist Party USA website, which defines liberty through a Marxist lens:

> Communists, on the other hand, see individuals in the context of their role in the economy...
> As long as property gives one person the right to control and subordinate another, real freedom isn't possible. So, we say that freedom requires abolishing class differences altogether: no more wage-workers, no more shareholders, just people working in a society based on collective, democratic control of resources.[12]

You see, communists always want to level the playing field of life at their own discretion. This bastardized version of liberty is rooted in envy, spite, and retribution, allowing the state outrageous power to control all levers of the economy, determine the rights of the people, employ mob rule, and pick winners and los-

---

12 Scott Hiley, "Communism, Freedom, and How We Get There," Communist Party USA, February 11, 2019, https://www.cpusa.org/interact_cpusa/communism-f reedom-and-how-we-get-there/.

ers. It also imagines a future utopia and what the Soviets referred to as "the new man," which seems identical to the wannabe world perceived by the architects of the climate agenda, namely the United Nations and the World Economic Forum.

No doubt, liberty, as defined by Locke and his contemporaries, is a delicate matter. This is why George Washington warned America's limited form of government could only survive if kept by a "moral and religious" people. Washington also insisted "virtue or morality is a necessary spring of popular government."[13]

Founding father and second president, John Adams, declared, "Public Virtue cannot exist in a Nation without private Virtue, and public Virtue is the only Foundation of Republics."[14]

Constructing a republic upon the principles of inalienable rights would be no easy thing, and those who signed the Declaration and carefully crafted our Constitution were not innocent dreamers; they were willing to risk their lives for this proposition. Though it would be a risky experiment, it was the best of all options.

Upon departing the Constitutional Convention of 1787, Benjamin Franklin was reportedly asked, "What kind of government shall we have, sir?"

Franklin replied, "A republic, if you can keep it."

THE MANIFESTO

Karl Marx and his followers were well aware of the representative, or republican, form of government established in America, and they loathed it. Marx perceived America's founders as reckless

---

13 George Washington, Farewell Address, September 17, 1796.
14 John Adams, Letter to Mercy Warren, April 16, 1776.

religionists peddling dangerous and preposterous propaganda—especially the inalienable rights noted in their Declaration of Independence.

To the Marxist, the *life* of an individual is not unique—just a fragment of the ever-multiplying collective mass—the result of a random, cosmic, Darwinian accident.

Likewise, *liberty* is an unattainable sentiment. Left to their own devices, the human masses are wholly incapable of coexistence without formidable government control and regulation.

And the *pursuit of happiness* through property ownership?

Preposterous.

Such professions were highly offensive to Marx. Free markets and capitalism, he assured, *always* result in a class struggle between the owners and controllers of the means of production (*bourgeoisie*) and the working class (*proletariat*). Marx preached the *proletariat* was oppressed and exploited by the *bourgeoisie*—period. He concluded that because capitalism seeks a profit, the *proletariat* suffers, because profit is made off the backs of their labor, and their labor causes the degeneration of their lives, health, and ability to enjoy life.

As an antidote to the presumptuous experiment being conducted in the United States, in 1849, Marx and Engels presented to the world their final formula for revolution, which they called the "Manifesto to the World." This infernal document would eventually be known as *The Communist Manifesto*.

In chapter two of their manifesto, Marx and Engels boldly state the goal of their envisioned new world order: "...the theory of the Communists may be summed up in the single sentence: abolition of private property."

As stated, property is not just physical possessions, it's also the stuff in your heart and soul. Personal property was a myth to Marx—*matter* was all there is, and no individual could claim matter as their own.

CONCLUSION

Hopefully, this opening chapter has been a primer on the two diametrically opposed political philosophies that have shaped the political universe in which we live. Before moving on, I would like to point out four significant differences between the communism/socialism of Marx and the liberty of Locke, espoused by America's founders:

- Marxism spares no room for a higher power, whereas liberty is based on such an assumption.

- Marxism perceives the life of an individual as part of the collective and therefore not unique; liberty seeks to protect life and one's ability to achieve happiness.

- Marxism allows for limited human rights promulgated by government; liberty demands natural rights for all humans and affirms the government upholds these rights.

Marxism's morality is relative and determined by the state; liberty's core values are derived from time-tested teachings handed down over the ages.

And then, there is the economy.

Devotees of Marx vehemently oppose capitalism and private enterprise, though they hypocritically allow their very elite allies to milk capitalism and pad their personal coffers.

On the other hand, free enterprise capitalism is an economic system involving trade and industry owned and controlled by individuals (businesses, corporations, shareholders, and the like) with the goal of garnering a profit for those same owners. Proceeds are then used freely by the individuals. Gains are used to purchase goods and services or for investment, thus, enhancing the vitality of the economy. Some save their wealth to prevent becoming a burden to society with hopes of providing an inheritance to future generations. Others give to the charities of their personal choice. These personal financial decisions have great potential to create robust economies. This is how America became the center of innovation, the breadbasket to the world, and home to the most charitable people in history.

But Marxists do not see it like that at all. They are driven by spite and retribution. For them, wealth distribution is unfair and unequal, therefore, private ownership of trade or industry is prohibited. Instead, profit-making enterprises must have common ownership, which is to say, they are owned by the community, which equates to the government. Profits are theoretically distributed equally among the people—but in all the communist countries to date, that has never happened. And charity? They see that as a responsibility of the government.

Theirs is a system completely contrary to Locke's assertion that, "Every man has a 'property' in his own person."

Marxism discourages hard work, constrains entrepreneurship, strips personal security, redefines charity, and routs *the pursuit of happiness*. Locke's philosophy of liberty encourages prosperity, peace of mind, and goodwill.

As a legal document, it is universally agreed that the Declaration of Independence was the constitutional means by which the founding fathers declared self-determination to the world. While severance from the British Crown was sought for many reasons (as detailed throughout the Declaration's grievances), the preamble explicitly defined the three distinct inalienable rights that became the basis for the Bill of Rights and the federal and various state constitutions.

For the last one hundred years, progressive legal scholars have attempted to methodically reinterpret the founders' intentions. One argument I noted in preparing this chapter states:

> [O]ne must keep the Declaration's text and meaning within the constraints of eighteenth-century legal thought, and remove all modern biases and political leanings. Perhaps the easiest argument by which one may dismiss the viewpoint that "life, liberty, and the pursuit of happiness" guarantees individualized natural rights or offers a presumption of liberty is to point out that the Declaration was an international document concerned with nations, not individuals.[15]

Obviously, wild mental gymnastics are required to recast the natural rights proclaimed by America's founders.

As you will see, Marx and some of his original followers were diabolically clever people who had more than one card up their

---

15 Patrick J. Charles, "Restoring 'Life, Liberty, and the Pursuit of Happiness' in Our Constitutional Jurisprudence: An Exercise in Legal History," *William and Mary Bill of Rights Journal*, Volume 20, Issue 2, December 2011.

sleeves to make sure communism spread, especially in America and the developed world: create an agenda that could be extraordinarily appealing to simple minds. To do this, they would use the environment. They would say over and over and over that the industrial revolution and mankind's insatiable lust for comfort has soiled the sky and compromised the climate.

And, like members of a well-programmed cult, the masses have bought in.

In the words of one of the planet's most influential environmental consortiums, Greenpeace, here is a simple summary of how climate change is being used to advance communism/Marxism on a sleeping world:

> Human rights express the entitlement of all people to be treated equally, to live their life in safety and freedom, and to be protected by their government. So many of our human rights, such as the right to life, health, food, and an adequate standard of living, are adversely affected by climate change. We see the evidence of this, for instance, with each new extreme weather event and the devastation that ensues, such as death and the destruction of crops and property. Without further action, climate change will continue to devastate people and the planet, and human rights will continue to be violated.[16]

---

16 "What Does Climate Change Have to Do with Human Rights?" Greenpeace International, December 10, 2018, https://www.greenpeace.org/international/story/19885/what-does-climate-change-have-to-do-with-human-rights/.

Greenpeace co-founder Patrick Moore resigned many years ago from his own organization after saying, "Greenpeace was 'hijacked' by the political left."

Moore went on to accurately explain, "The 'environmental' movement has become more of a political movement than an environmental movement. They are primarily focused on creating narratives, stories, that are designed to instill fear and guilt into the public so the public will send them money."[17]

You can't get any plainer than that.

17 "Greenpeace Founder Patrick Moore Says Climate Change Based on False Narratives," Frontier Centre for Public Policy, September 16, 2022, https://fcpp.org/2022/09/16/greenpeace-founder-patrick-moore-says-climate-change-based-on-false-narratives/.

44

CHAPTER TWO

EARTH GETS HER DAY

APRIL 22, 1970. THE INAUGURAL "celebration" of a brand-new holiday called Earth Day. And what kid wouldn't want another excuse to celebrate?

Smiling like it was Christmas, my eighth-grade earth science teacher walked around the room distributing green Earth Day buttons, mandating that all of his classes would be "celebrating" this new holiday. He mentioned something about it being a "teach-in" and urged us to proudly pin the buttons on our shirts so we could do our part in "saving the earth." That *had* to equate to a good grade in class, we figured. So, we all happily complied.

Heck, with all the environmental disasters we had been seeing on the nightly news, who wasn't for clean water and clean air?

But that evening at the dinner table, my dad expressed his thoughts and views on the new national holiday in his usual direct and eloquent style.

"Bullshit!"

No, sir. They weren't pulling one over on *his* hardworking, practical generation. If they smelled a scam, they had absolutely no problem warning folks not to step in it.

"So, we celebrate the earth now—like a birthday party?" he said.

Oh, could his words drip with sarcasm. I just shrugged, embarrassed.

"This is what I'm paying my taxes for?"

"My science teacher handed out the buttons."

"And you just pinned it on? Without questioning it?"

I felt like the biggest goof.

"He said we'd be helping save the earth."

"Save the earth? Really? Like Superman? How 'bout I call you Super Brian?"

"No," I mumbled, shrinking in my chair. "We just wear the button on T-shirts to remind people to take care of the environment."

"Alice," he said, turning to my mom. "Did you hear that? The earth doesn't know how to take care of itself anymore."

I was dying to get out of the conversation. And mercifully, my dad let me.

"You know, I was thinking. Maybe I should be a nice guy and buy a gift for Mother Earth. But then I thought, what can you buy for somebody who has *everything*?"

I was afraid to look up, but when I did, I saw that gleam in his eye and I cracked up.

He joined in, slapping me on the shoulder with pride that I finally understood his point. I removed the moronic button from my shirt and tossed it on the table.

Dad picked it up, chuckling. "Can you believe this, Alice?"

Nonetheless, it was, and still is, an amazingly devious scheme—utilizing nature as a highly effective method to justify a prejudice against capitalism, liberty, and human contentment. But the idea has been around since at least 1862.

Professor Justus von Liebig was a German chemist and a colleague of Karl Marx's who shared Marx's views on materialism and communism. He published an updated version of an otherwise mundane book he wrote twenty-two years prior entitled *Organic Chemistry in Its Application to Agriculture and Physiology.* The new edition was unique in that for the first time, a scientist used his lectern to advance an environmental argument attacking free markets regarding bird guano—bird droppings.

In the mid-1800s, British citizens were living longer and healthier lives in comparison to the rest of the world. Much of this good life was the result of newly developed domestic farming techniques capable of delivering an abundance of affordable food to the people. One of the key ingredients in the British farming success was the use of guano, a very efficient fertilizer. Farmers were willing to purchase the ordure from anyone who would sell it at a reasonable price because, quite simply, it worked. Guano imports to England began in 1841, and twenty years later, it is estimated some 3.2 million tons of the phosphate-rich additive had been imported into the country.[1] Guano was being brought to market from mountaintops, fields, and caves in Europe, North America, South America, Africa, and the Caribbean Islands.

Though von Liebig well understood the theoretical benefits of employing guano as a fertilizer, he held significant disapproval for several reasons. First, it was his opinion that while collecting deposits of the organic material, workers were ruining surrounding environs. Second, von Liebig contended that guano traders were exceedingly greedy and taking advantage of underpaid work-

---

1   C. C. Hoyer Millar, *Florida, South Carolina, and Canadian Phosphates* (London: Eden Fisher and Co., 1892), 15.

ers to turn a profit. Third, he was riled that the crops benefiting from guano were growing at a rate that he felt superseded nature's intention; the increased yields created more vegetables and meat for Londoners to eat and more feed for livestock. Yes, von Liebig would admit, the people were now living longer, healthier lives and tended to have larger families, but the escalation in prosperity and family size had led to larger houses and more animals. The continual progression would demand more food, more feed, and more excrement and pollution.

Von Liebig described guano as being at the center of a "robbery system." Using variegated imagery, von Liebig said Great Britain's use of guano,

> ...deprives all countries of the conditions of their fertility. It has raked up the battlefields of Leipsic, Waterloo, and the Crimea; it has consumed the bones of many generations accumulated in the catacombs of Sicily; and now annually destroys the food for a future generation of three millions and a half of people. Like a vampire it hangs on the breast of Europe, and even the world, sucking its lifeblood without any real necessity or permanent gain for itself.[2]

For von Liebig, there was no compromising solution—in fact, there never is for those who hold tight to such philosophies.

---

2   The translation of this passage from the introduction to the 1862 edition of von Liebig's book is from Erland Marold's "Everything Circulates: Agriculture, Chemistry, and Recycling Theories in the Second Half of the Nineteenth Century," *Environment and History* 8 (2002), 74.

Justus von Liebig is the first recorded fellow traveler with a physical science PhD to attack capitalism based on environmental standards. His shrill strategy was thoroughly vetted by Marx, eventually becoming the lever of choice for future socialists and communists who, like a henhouse full of Chicken Littles, desperately try to convince the world the sky is falling—and those seeking a better life are guilty of destroying nature.

MARX'S MALICE

At the time of Dr. von Liebig's launch on Great Britain's agricultural methods, Marx was in the process of completing one of his signature works, *Das Kapital.* He was quite impacted by von Liebig's complex book on organic chemistry and in an 1866 letter to Engels wrote, "I had to plough through the new agricultural chemistry in Germany, in particular von Liebig…which is more important for this matter than all of the economists put together."[3]

In *Das Kapital,* Marx espoused what he commonly referred to as "natural wealth," which he described as "fruitful soil, waters teeming with fish, etc., and…waterfalls, navigable rivers, wood, metal, coal, etc."[4]

Marx, like von Liebig, was convinced such natural wealth did not belong to man and could only be utilized if necessary for the absolute common good and without anyone garnishing a profit along the way.

---

3   Letter from Marx to Engels, February 13, 1866, from Marx and Engels, *Selected Correspondence* (New York: International Publishers, 1942), 204–205.

4   Karl Marx, *Das Kapital,* Volume 1, Part V, Chapter Sixteen, "Absolute and Relative Surplus Value," https://www.marxists.org/archive/marx/works/1867-c1/ch16.htm.

Marx went on to state that one of "von Liebig's immortal merits" is having "…developed from the point of view of natural science, the negative, i.e., destructive, side of modern agriculture."[5] Regarding von Liebig's extreme criticism of the guano trade, Marx focused less on the cycle of bird secretions and more on the economics of labor and his biased perception of modern farming:

- British traders extracted natural wealth to garner a personal profit. In the long process of bringing guano to market, lower-class laborers were continually exploited.

- As crop yields expanded, natural wealth was further overburdened as landowning farmers gained higher profit margins.

- The resulting bountiful crops provided an overabundance of feed for the livestock, which further enabled the farmers to reap increased profits by raising more animals at lesser cost.

Von Liebig and Marx perceived a chain reaction that they loathed: city dwellers were now able to purchase more food at reduced prices, encouraging them to have larger families. Larger families required bigger houses to be built by more exploited workers. The urban population boom required more horses for transportation and more subsequent animal dung, which had to be removed from the cities and hauled to the dump by laborers they ascribed as overworked, underpaid, and exploited.

---

5   Karl Marx, *Das Kapital*, Volume 3, Part I, Chapter Six, "The Effect of Price Fluctuation," Section Two, http://www.marxists.org/archive/marx/works/1894-c3/ch06.htm.

Marx saw all of this as a confirmation of his laws of matter. He described the guano trade as a vicious cycle, perpetuated by a lust for profit, writing in *Das Kapital*, "[T]he increased exploitation of natural wealth by the mere increase in the tension of labor-power, science and technology give capital a power of expansion."[6]

In the mind of Marx, a capital-based system of agriculture was irrational. "The moral of history," he said, "is that the capitalist system works against a rational agriculture, [and] that a rational agriculture is incompatible with the capitalist system."[7]

He further opined:

> ...all progress in capitalistic agriculture is a progress in the art, not only of robbing the laborer, but of robbing the soil; all progress in increasing the fertility of the soil for a given time, is a progress towards ruining the lasting sources of that fertility. The more a country starts its development on the foundation of modern industry, like the United States, for example, the more rapid is this process of destruction.[8]

---

6   Karl Marx, *Das Kapital*, Volume 1, Chapter Twenty-Four, "Conversion of Surplus-Value into Capital," Section 4, http://www.marxists.org/archive/marx/works/1867-c1/ch24.htm.

7   Karl Marx, *Das Kapital*, Volume 3, Part I, Chapter Six, "The Effect of Price Fluctuation," Section 2, http://www.marxists.org/archive/marx/works/1894-c3/ch06.htm.

8   Karl Marx, *Das Kapital*, Volume 1, Chapter Fifteen, "Machinery and Modern Industry," Section 10, http://www.marxists.org/archive/marx/works/1867-c1/ch15.htm#a245.

Karl Marx was a pathetic man. His entire life's work was built upon a foundation of anger and bitterness. He held a distinct grudge against those who pursued a personal course of happiness and, obviously, had an ax to grind with the representative republic of the United States. Whether human contentment was found in faith, financial wealth, or the work of one's own hands, Marx was consumed with malice. Such anger is consistent with so many on the progressive side of the aisle today.

Historian Paul Johnson provides an illuminating glimpse into Marx's personal life in his fascinating book, *Intellectuals*:

> In all his researches into the iniquities of British capitalism, [Marx] came across many instances of low-paid workers but he never succeeded in unearthing one who was paid literally no wages at all. Yet such a worker did exist, in his own household.... This was Helen Demuth [the lifelong family maid]. She got her keep but was paid nothing.... She was a ferociously hard worker, not only cleaning and scrubbing, but managing the family budget.... Marx never paid her a penny....
>
> In 1849–50...[Helen] became Marx's mistress and conceived a child.... Marx refused to acknowledge his responsibility, then or ever, and flatly denied the rumors that he was the father.... [The son] was put out to be fostered by a working-class family called Lewis but allowed to visit the Marx household [to see his mother]. He was,

however, forbidden to use the front door and obliged to see his mother only in the kitchen.[9]

The hostility directed at economic achievement described in *Das Kapital* is still held in high regard by progressives today: in their worldview, no one has the right to profit from the sale and distribution of natural resources such as food, water, timber, coal, gas, or oil. And, whether it's saving the forests, whales, snails, or the climate, it all comes back to a deep-rooted contention that the quest for such profit is always underhanded. They are convinced free enterprise will ultimately destroy the planet.

FOUNDING THE AGENDA

Beyond Marx and von Liebig, there are three additional founders of the green agenda who need to be brought to light, as they are revered by environmentalist leaders and teachers today.

Sir Edwin Ray Lankester (1847–1929) was a zoologist at University College, London. Though some thirty years younger than Marx, the two were close friends, colleagues, fellow materialists, and communists. Lankester was a frequent guest at Marx's home in the last few years of Marx's life and attended his funeral.

Lankester is noted as the greatest Darwinist of his generation—in fact, it is well established that Lankester's family was friends with Charles Darwin, and much has been written of Lankester being "carried on the shoulders of Darwin" as a child.[10]

Regarding *Das Kapital*, Lankester once penned a pun to Marx, clearly taking a swipe at America's inalienable rights, say-

---

9   Paul Johnson, *Intellectuals* (Harper and Row Publishers, 1988), 79–80.

10   John Bellamy Foster, "Marx Ecology in Historical Perspective," *International Socialist Journal*, Issue 96, Winter 2002.

ing that he was absorbing "your great work on *Kapital*...with the greatest pleasure and profit."[11]

Lankester was the most prolific eco-socialist thinker of his time, writing powerful papers indignant of human behavior with an urgency unprecedented until the late twentieth century. In Lankester's most popular screed, *Nature and Man*, he describes humans as the "insurgent son" of nature.[12]

According to Lankester,

> We may indeed compare civilized man to a successful rebel against nature who by every step forward renders himself liable to greater and greater penalties...[H]e has willingly abrogated, in many important respects, the laws of his mother Nature by which the kingdom was hitherto governed; he has gained some power and advantage by so doing, but is threatened on every hand by dangers and disasters hitherto restrained: no retreat is possible—his only hope is to control...the sources of these dangers and disasters.[13]

Lankester's star pupil was Arthur Tansley, a foremost academician specializing in botany, and the man noted for coining the term "ecosystem." Born in 1871, Tansley personally never interfaced with Marx but was a fellow Darwinist, materialist, socialist who was teeming with animosity towards humankind. Tansley said he was deeply concerned with "the destructive human activ-

---

11   Ibid.
12   Sir Edwin Ray Lankester, *Nature and Man* (Oxford Clarendon Press, 1905), 23.
13   Ibid., 27.

ities of the modern world." He argued, "Ecology must be applied to conditions brought about by human activity."[14]

In the 1940s, Tansley had a young protégé named Charles Elton who worked with him to further develop the ecosystem concept, particularly "invasion ecology" whereby non-native species are hereby introduced into a region by capitalists. His fiery writing style set the stage for today's eco-authors, with quotes like this, from a book published in 1958:

> It is not just nuclear bombs and war that threatens us. There are other sorts of explosions, and this book is about ecological explosions.[15]

From Karl Marx to Charles Elton, there are a mere three degrees of separation bringing us to the modern, radical environmental movement. However, there is another key historical figure who must be properly highlighted; he was the first political leader to implement the green agenda.

And his plan did not end well.

GREEN MONSTER

Vladimir Ilyich Lenin.

Mention the name to any US citizen formerly from the Soviet Union, and the response will be instant and visceral. Lenin was the Marx-honoring communist who overthrew Russia and birthed a movement of tyranny, eventually plunging Russia and Eastern Europe into generations of doom and misery.

---

14 Sir Arthur G. Tansley, "The Use and Abuse of Vegetational Concepts and Terms," *Ecology*, vol. 16, no. 3 (July 1935), 299, 303–304.

15 Charles Elton, *The Ecology of Invasions by Animals and Plants* (London: Methuen and Co., 1958), opening paragraph.

BRIAN SUSSMAN

Lenin was born in 1870 into a family steeped in revolutionary thought. When he was seventeen, his older brother was executed for attempting to assassinate the czar. Several years later, Lenin began to engross himself in the works of Marx. By the early 1900s, he was a well-known Marxist author writing books on materialism and socialist economic theory. In 1916, he authored a furious missive, *Imperialism, the Highest Stage of Capitalism*. Having gained a significant following by October of the following year, he and a small band of cohorts, not without outside international help, staged a cunning coup. Lenin was named chairman of the new government, and the Russian Revolution had begun.

Immediately, members of the former regime were arrested and, in many cases, executed. Banks were quickly nationalized, private businesses taken over by the state, and a supreme economic council formed to run the economy. All private land, including any property belonging to the church, was now that of the new Soviet state. Civil war ensued as freedom fighters attempted to withstand the new government and its Red Army, but they were eventually brutally defeated. Estimates vary as to the number of deaths during this time of upheaval, but between men killed in action, those executed by the Red Army, civilian casualties, and those who perished of exposure and disease during Lenin's seven-year reign, the number is thought to be in the millions. Russia's economy was devastated by the war, with factories and infrastructure destroyed, livestock and raw materials pillaged, mines flooded, and the people lacking food, shelter, and hope. Forced labor camps and police-state terror became the tools of comprehensive order.

And yet, despite the vast human tragedy occurring under his watch, one of Lenin's top priorities from the very beginning of his dictatorship was to institute a comprehensive green agenda. It is mind-boggling to consider how much time this madman must have spent on hammering out the details of this exhaustive plan.

Besides being a devout student of Marx, Lenin was quite familiar with von Liebig and Lankester and, like them, believed that nature's resources should never be used for profit of any sort—only for the common good, and even then, only if absolutely necessary.

In 1918, within his first year as party chairman, Lenin issued a policy paper entitled "Decree on Land," declaring all forests, waters, and minerals property of the state. Later that same year, as locals began to desperately harvest portions of the forest for food, firewood, and construction material, Lenin issued another lengthy diatribe entitled "Decree on Forests." From that moment, the forests were protected, and only certain small, insignificant sectors were established for use. Lenin's mandate declared the protected areas as a "preservation of monuments of nature."

Animal rights came next with another protracted edict, "Decree on Hunting Seasons and the Right to Possess Hunting Weapons," which went into force in May 1919. It banned the hunting of game, such as moose and wild goats, and ended the open seasons for a variety of other animals in spring and summer. No surprise that starvation ensued.

Lenin's counsel in crafting this green agenda came from acclaimed Russian agronomist N. N. Podiapolski, who urged the immediate creation of *zapovedniki*, or nature preserves. In such preserves, nature would be completely free from humans—no hunting, harvesting, clearing of dead growth, mowing, sowing, or even the gathering of fruit. All activity in such regions was illegal. Podiapolski recalled one meeting with Lenin, which was convened despite the chairman's involvement in a fierce military campaign:

> Having asked me some questions about the military and political situation in the Astrakhan region [a fertile area rich in natural resources located in southwestern Russia on the delta of the Volga River, sixty miles from the Caspian Sea], Vladimir Ilyich expressed his approval for all of our initiatives and in particular the one concerning the project for the *zapovednik*. He stated that the cause of conservation was important not only for the Astrakhan region, but for the whole republic as well.[16]

Podiapolski drafted a resolution soon approved by the Soviet government in September 1921 entitled "On the Protection of Nature, Gardens, and Parks." A commission was established to oversee implementation of the new laws. One of the first tasks was to create another *zapovedniki* on the slopes of the South Ural Mountains, an area rich in coal, iron ore, non-ferrous metals, and gold. Despite the enormous potential economic value to the state from the minerals, Lenin felt the region had much to offer in revealing evolutionary geological processes.

During Lenin's horrific political experiment as the supreme Soviet leader, he initiated a most audacious nature conservancy

---

16 Douglas Weiner, *Models of Nature: Ecology, Conservation and Cultural Revolution in Soviet Russia* (Pittsburgh, PA: University of Pittsburgh, 2000), 27.

program, showing immense grace towards nature and utter contempt for his fellow man. Originating with a vision cast by Marx fifty years prior, Lenin had successfully implemented the green agenda's version one.

His accomplishments would eventually be celebrated the world over.

BESTSELLERS

The stage had been set for presenting ecological issues to smear capitalism and especially American liberty. In the sixties, two cunning authors, Rachel Carson and Paul Ehrlich, independently put forth tomes that mixed seductive cocktails of ecology, sociology, and political science. Though their books became bestsellers, the content is peppered with statistics and statements that ultimately could not withstand the microscope of truth.

Carson's work came first in 1962 with the publishing of *Silent Spring*. She held a degree in zoology and found employment creating brochures for the United States Department of Fish and Wildlife from 1936 to 1950. Longing to become a successful author, Carson's first effort fell flat, but in her 1951 second book, *The Sea Around Us*, she proved herself to be a gifted word stylist. A decade later, she released her opus, *Silent Spring*. A close examination of the book, however, reveals how Carson's work was heavily influenced by a cadre of Marxists.

One such associate was H. J. Muller, a Nobel Prize recipient in genetics. Throughout the pages of *Silent Spring*, Carson makes go-to references to much of Muller's academic work. She failed, though, to mention that Muller was a known anti-American socialist who had become so disenchanted with the US that he moved to Nazi Germany in 1932 and eventually to the Soviet Union.

Carson was also obsessed by the work of Tansley and Elton, even twice borrowing a line from Elton—*rain of death*[17]—in an April 1959 letter to the *New York Times* in which she introduced her attack on pesticides and again in *Silent Spring*'s chapter, "Indiscriminately from the Skies."[18]

A fourth significant influence on Carson's thinking was her friend Robert Rudd, a professor of zoology at the University of California, Davis. Rudd is described by a noted socialist historian as "a sophisticated left thinker with a deep sense of the ecology, sociology, and political economy...."[19]

Carson first contacted Rudd in April 1958 to receive assistance in writing *Silent Spring*, and they developed a strong friendship and a close working relationship.[20] Carson drew extensively on Rudd's research in two *Silent Spring* chapters morosely titled "And No Birds Sang" and "Rivers of Death."

For Carson, ecology emerged as the basis for a radical challenge to human development, which she unmistakably loathed. "The modern world," she declared, "worships the gods of speed and quantity, and of the quick and easy profit, and out of this idolatry, monstrous evils have arisen."[21]

---

17 Charles Elton, *The Ecology of Invasions by Animals and Plants* (London: Methuen and Co., 1958), 137–142.

18 Rachel Carson, *Silent Spring* (Boston: Houghton Mifflin Company, 1962), 155.

19 John Bellamy Foster, "Rachel Carson's Ecological Critique," *Monthly Review*, January 2008.

20 John Bellamy Foster and Brett Clark, "Rachel Carson's Ecological Critique," *Monthly Review*, February, 2008.

21 Rachel Carson, "Foreword," in Ruth Harrison, *Animal Machines* (Vincent Stuart Ltd, 1964), vii.

In a clarion call for the formation of a global environmental movement to combat modern development, Carson proclaimed, "The struggle against the massed might of industry is too big for one or two individuals...to handle."

Borrowing from Marx's laws of matter in a rare television interview, Carson stated that "man's endeavors to control nature by his powers to alter and to destroy would inevitably evolve into a war against himself, a war he would lose unless he came to terms with nature."[22]

ONE-TWO PUNCH

A few years after Carson's book stole the minds of many, a second potent work was published in 1968, Paul Ehrlich's *Population Bomb*. Ehrlich, a professor emeritus at Stanford University, has authored many bestselling social engineering books over the decades, but *Population Bomb* was his first hit. Like *Silent Spring*, it became required reading in many public schools in the early seventies. "The battle to feed humanity is over," Ehrlich falsely prophesied. "In the 1970s and 1980s hundreds of millions of people will starve to death in spite of any crash programs embarked upon now."[23]

Ehrlich's anti-human message was a hybrid of radical thought, seemingly drawing from concepts articulated in Marx's laws of matter, von Liebig's guano argument, Lenin's radical conservation program, and Thomas Robert Malthus's *An Essay on the Principle of Population*, written in 1798.

---

22 United States Forest Service, "Rachel Carson National Wildlife Refuge," https://www.fws.gov/staff-profile/rachel-carson-1907-1964-author-modern-environmental-movement..

23 Paul Ehrlich, *The Population Bomb* (New York: Ballantine Books, 1968), 1.

Malthus believed that unchecked population growth always exceeds the food supply. He contended that improving the lives of the lower classes or improving agricultural conditions was fruitless because these steps would only encourage humans to have more offspring, which would exacerbate the original problem and prevent society from "perfectibility."

Ehrlich has long opined that the earth is being forced to support too many people who require too many resources and who produce too much pollution. His final solution has always been clear: "Population control is the only answer."[24]

Ehrlich's wild assertions in *Population Bomb* include equating people with a cancer that must be eradicated:

> A cancer is an uncontrolled multiplication of cells; the population explosion is an uncontrolled multiplication of people.... We must shift our efforts from treatment of the symptoms to the cutting out of the cancer. The operation will demand many apparently brutal and heartless decisions.[25]

The method to Ehrlich's madness was revealed in his action plan:

> Our position requires that we take immediate action at home and promote effective action worldwide. We must have population control at home, hopefully through changes in our

24   Ibid.
25   Ibid., 166.

value system, but by compulsion if voluntary methods fail.[26]

While Carson convinced a generation that modern American liberty, ingenuity, free markets, and capitalism were the problems ruining the planet, Ehrlich introduced the solution: a change in our value systems…by compulsion if other methods fail.

Marx and Engels could not have executed a more effective one-two punch.

POLLUTION ON PARADE

As if on cue, a series of ecological mishaps hit the news, anointing Carson and Ehrlich as messengers of ecological enlightenment.

The New York City garbage collectors' strike of 1968 was hailed as the greatest ecological disaster of the time and hyped by those on the left as undeniable proof that humankind was trashing the planet. With only three television network newscasts to choose from at the time, producers, who understood that alarming stories kept viewers fixed to the set, made sure that every American witnessed trash piling up on New York's sidewalks, which they told us measured "10,000 tons per day."[27]

Pollution was on parade.

In January of 1969, a Union Oil drilling platform six miles off the coast of Santa Barbara, California, sprang a leak that allowed hundreds of thousands of gallons of crude oil to seep into the Pacific and wash ashore. Never mind the fact that long before the drilling platforms had been erected, Santa Barbara beaches were

26 Ibid., xi–xii.
27 H. Lanier Hickman Jr., *American Alchemy: A History of Solid Waste Management in the United States* (Forester Press, 2003), 520.

well-known for small, sticky balls of tar strewn everywhere across the sand. There is so much oil naturally leaking off the Pacific coastline that it literally bubbles up from the ocean floor!

Nonetheless, the news media provided activists with yet another convenient tool to demonize the American lifestyle, as oil-coated birds were stuck in the same muck used to power America's growing fleet of automobiles. The nation's first outspoken congressional environmentalist, Wisconsin senator Gaylord Nelson, immediately flew to California to see the action for himself. He returned to Washington, angered at the oil industry, vowing to "get the nation to wake up and pay attention to the most important challenge the human species faces on the planet."[28]

Several months later, in June of 1969, another event was etched into the American psyche when the Cuyahoga River in Cleveland, Ohio, burst into another symbol of a planet in disrepair.

The popular story says the entire river was consumed in flames and burned for hours, but the truth is the fire burned for less than thirty minutes, and no conclusive evidence of its cause has ever been determined, though it is widely accepted that the combination of industrial waste and floating debris somehow ignited beneath a train trestle.

The blaze was extinguished so quickly that nary a photographer had time to snap a photo, and the local media instead had to settle for a fireboat hosing down the charred trestle. However, the lack of visual evidence did not stop *TIME* magazine from

---

28 "Gaylord Nelson Promotes the First Earth Day," United States Senate, April 22, 1970, https://www.senate.gov/artandhistory/history/minute/Gaylord_Nelson_Promotes_the_First_Earth_Day.htm.

running a dramatic cover shot of the Cuyahoga River aflame in their next edition.

However, the grossly misleading photo was taken during a much more serious fire that occurred on the river in 1952.

The ecological news events of '68 and '69 coincided with hundreds of thousands of hippies and young Marxist revolutionaries taking to the streets of America to protest the Vietnam War, specifically, and capitalism at large. These were the biggest protests ever witnessed in the US, and the nightly news broadcasts brought the drama into every living room in the country. A movement had been born, fueled by new narcotics, new music, and a new desire to "get back to the garden," as many were fond of saying.

The defining moment occurred on a farm in Woodstock, New York, in August of '69. The Woodstock Music and Art Fair drew two hundred thousand young people who camped in the rain and mud for three days of peace, love, drugs, rock and roll, and the launch of a radical ecological vision.

MOTHER EARTH GETS HER DAY

Seizing the Woodstock moment, Senator Nelson recruited Ehrlich to be part of a national environmental teach-in steering committee designed to unleash a grassroots response to the recent ecological news stories.[29]

I recall these "teach-ins." They were ridiculous. Scrapping the assigned curriculum for the day, teachers would have students sit cross-legged on the floor and "rap" about how America was

---

29 "A Grassroots Movement," Gaylord Nelson, Founder of Earth Day, https://nelsonearthday.net/grassroots-movement/.

an imperialist nation and why socialism really wasn't such a bad form of government—it just needed to be implemented properly.

Nelson's efforts were aided by a young liberal activist named Denis Hayes. Hayes was a former student body president from Stanford with an effective track record for organizing anti-war, anti-America protests. While pursuing a master's degree in public policy at Harvard, Hayes had heard about the teach-in concept and sought out Nelson to help him take his strategy of infiltrating the classroom nationwide.[30]

"My God," Nelson said, following his meeting with Ehrlich, "why not a national teach-in on the environment?"[31]

Years later, Nelson elaborated, "I was satisfied that if we could tap into the environmental concerns of the general public and infuse the student anti-war energy into the environmental cause, we could generate a demonstration that would force this issue onto the political agenda."[32]

Soon, Senator Nelson formally announced there would be a "national environmental teach-in" on the crisis of the environment in the spring of 1970.[33] Ehrlich and Hayes would play pivotal roles—Ehrlich as an academic influencer and Hayes as the rebellious organizer. After careful consideration, a name and date for the event were chosen: Earth Day, April 22.

---

30 Christina Pazzanese, "How Earth Day gave birth to environmental movement," The Harvard Gazette, April 17, 2020, https://news.harvard.edu/gazette/story/2020/04/denis-hayes-one-of-earth-days-founders-50-years-ago-reflects/.

31 Ibid.

32 "How the First Earth Day Came About," Senator Gaylord Nelson, *American Heritage Magazine*, October, 1993.

33 "Tracing Earth Day's Origins," Gaylord Nelson, Founder of Earth Day, https://nelsonearthday.net/gaylord-nelson-earth-day-origins/#proposal.

CONCLUSION

Many times, while speaking to audiences across the country, I have asked, "How many of you are originally from the former Soviet Union?"

Hands always go up. I then ask those former Soviets, "What is the significance of April twenty-second?"

With scowls and derision, they immediately reply in unison, "Vladimir Lenin's birthday!"

On April 22, 1970, Lenin would have been one hundred years old. Though Senator Nelson denied it, it would seem quite plausible that selecting Lenin's date of birth to "celebrate" Earth Day was purposeful. In light of Lenin's unrestricted devotion to nature for his own evil purposes and at the expense of his own people, this highly revered communist vanguard provided the perfect model for a government's environmental agenda.

In addition, Lenin also perceived government-run education as the most efficient means of indoctrination. An infamous quote attributed to him states, "Give us the child for eight years, and it will be a Bolshevik forever."

Birthing the modern environmental movement in the public schools and universities, along with so many other godless socialist ideas, seems to have been the perfect scheme.

An honest examination of Earth Day reveals the fact that it has never been a celebration about the beauty and bounty of this awesome world we have been blessed with but, instead, has always been an increasingly Marxist assault on humankind.

By now, generations have been deeply indoctrinated, according to plan.

During the first decade of Earth Day observances, humans were named polluters. By the eighties, the event's organizers recast mankind as tree killers, and in the early nineties, humanity was christened the species annihilators. Global warming and climate change scares never really became mainstream until the mid-nineties, but when they did, they provided compatriots at Earth Day headquarters, along with their malevolent fellow progressives, with the ultimate propaganda stratagem to herd society toward a global socialist reset. Human greed and counterfeit liberty have become the cult's latest fear pornography, promising to rapidly destroy the planet's atmosphere and placing every single person at risk. Except them, of course.

A variable often overlooked or disregarded in this communist cult movement is their generational tenacity to seek and gain power at all costs over every institution and every living soul. With things moving at the speed of light on societal, scientific, technological, and political levels, we can be sure that we will see even more absurd sleight of hand from madmen and madwomen in order to maintain power over the lives of all humanity.

Looking back, I'd have to conclude that my dad called it on the money.

We're in a battle, and we can't stop fighting. Thanks, Dad.

CHAPTER THREE

DESIGNER POLLUTANT

THIRTEEN MONTHS BEFORE THE FIRST Earth Day in 1970, the newly inaugurated president of the United States, Richard Nixon, visited the beaches of Santa Barbara for a firsthand look at the aforementioned oil spill.

The ecological emergency occurred in federal waters on January 28, 1969, when engineers working on a new Union Oil platform began pulling a drilling tube from the ocean's floor to replace a drill bit. While the water was less than two hundred feet deep, the drilling tube extended over half a mile—3,500 feet. During the routine procedure, a pressure differential unexpectedly occurred resulting in a "blowout," with natural gas, oil, and mud shooting up from the well into the ocean.

Union Oil had four operational oil platforms in the area, and this was to be their fifth. In accordance with federal regulations, the platforms required three hundred feet of protective steel casing to extend below the ocean floor, however, engineers from Union Oil believed 239 feet would be sufficient and received an exemption from President Lyndon Johnson's Department of Interior. Speculation has always been that another sixty-one feet of casing could have prevented the blowout.

It took eleven days to properly cap the spill, during which time it is estimated that up to one hundred thousand barrels of crude oil were released into the ocean and on to the adjacent shore.

There is no getting around it; the initial environmental impact was horrible, as the ocean and beaches were covered in black tar.

The mess made for compelling television as the nation watched volunteers spread enormous piles of straw over oiled sections of the beach, raking it into gooey piles. County and state workers used steam to clean the oil off beachside boulders. Planes were employed to drop chemical dispersants to help break up the oil, even though those chemicals were eventually proven to be toxic to wildlife.

As a former governor of California, President Nixon knew Santa Barbara well. And, as a keen politician, he also could read the tea leaves: the new, popular environmental movement had potential civic clout. On March 21, with great fanfare, he cruised the skies above the coast to observe the scene for himself and then walked the beach in his shiny leather oxfords surrounded by a gaggle of reporters. While the white sand had been remarkably cleaned, he found a few random tar balls to nudge with his toe.

A hundred yards away, a crowd of protesters chanted, "Get oil out! Get oil out!"

"I don't think we have paid enough attention to this," Nixon said to the reporters tightly gathered about him. "We are going to do a better job than we have done in the past."[1]

---

1   Meir Rinde, *Distillations Magazine*, "Richard Nixon and the Rise of American Environmentalism," Science History Institute, June 2, 2017, https://www.sciencehistory.org/distillations/richard-nixon-and-the-rise-of-american-environmentalism.

Ecology activists were skeptical of the new president's assurances and rightly so. Nixon had no record on the environment and never highlighted the issue during his campaign. But he would soon propose an ambitious and expensive pollution-fighting agenda to Congress and would even go on to create the Environmental Protection Agency (EPA). Working through a reorganization plan rather than legislation, Nixon consolidated functions that were scattered among forty-four government offices, declaring the EPA would treat "air pollution, water pollution, and solid wastes as different forms of a single problem."[2]

Next came the federal Clean Air Act, which gave the EPA authority to set standards for outdoor air quality and the amount of pollution individual industries could emit. The measures also established enforceable emission standards for new cars. Nixon, a proponent of state's rights, easily justified his ecological efforts because air and water traverse state lines.

He championed his environmental record during his successful run for a second term, and in an address to the legislature in February 1973, the president declared the air pollution crisis over. "I can report to Congress that we are well on our way to winning the war with environmental degradation, well on our way to making peace with nature."[3]

In many ways, Nixon was spot on.

In the 1960s, the smog in Los Angeles was so bad, one could barely see the seven-thousand-foot peaks of the nearby San Gabriel mountains. I vividly recall my dad rescuing me from a swimming pool because of a smog-induced asthma attack when

---

2   Ibid.
3   Ibid.

I was a kid. Air quality was so bad in the nation's big cities like LA, Chicago, Detroit, and New York that sometimes taking a deep breath while exercising made one's lungs feel as if they were on fire.

All these years later—with millions more residents across America and countless more vehicles, houses, and businesses—in large measure, thanks to the Clean Air Act, nearly 70 percent of the total emissions of the six major pollutants that were once fouling the sky—carbon monoxide, lead, ground-level ozone, nitrogen dioxide, particulate matter, and sulfur dioxide—have all notably decreased.[4]

America's spacious skies are not filthy like they once were. Unlike other countries (China and India are among today's worst offenders), the United States has done an incredible job of cleaning up the atmosphere.

But one element that was glaringly missing on the list of sky-high poisons was a chemical compound that was, even several decades ago, never considered a threat to humanity: carbon dioxide (CO_2), the designer pollutant.

---

4   Section 109 of the Clean Air Act requires the EPA to establish National Ambient Air Quality Standards (NAAQS) for air pollutants that endanger public health or welfare, in the administrator's judgment, and whose presence in ambient air results from numerous or diverse sources. Using this authority, the EPA has promulgated NAAQS for six air pollutants or groups of pollutants: sulfur dioxide (SO_2), particulate matter (PM2.5 and PM10), nitrogen dioxide (NO_2), carbon monoxide (CO), ozone, and lead. The act requires the EPA to review the scientific data upon which the standards are based every five years and revise the standards if necessary. More often than not, the EPA has taken more than five years in reviewing the standards, but the establishment of a deadline has allowed interested parties to force review of the standards by filing suit.

CO_2-101

World governments are implementing ambitious and expensive plans to limit the increase of global temperatures by greatly reducing CO_2 emissions to achieve what they refer to as a "net zero" carbon economy in keeping with their argument that carbon emissions are the primary driver of global temperature increases. Their plans would radically alter the established way of life for everyone, particularly in the United States, which generates more carbon dioxide per capita than any other nation.

It is crucial to understand that these schemes are built on science that, at best, is far from settled. In fact, it is deceptively rigged.

A quick little lesson for the nonexpert—a carbon dioxide molecule is made up of one carbon atom and two oxygen atoms—hence, the designation, CO_2. Carbon dioxide and oxygen are essential to all life. As you're reading this, you're inhaling oxygen and exhaling CO_2, while our plants and trees are absorbing CO_2 and emitting oxygen.[5] It's an amazingly efficient life-supporting system.

Carbon is the chemical backbone of all living organisms, from plankton to people. When new life is created, carbon forms key molecules like protein and DNA. It's also important to note that all of the carbon we have on earth is the same amount we've

---

5   Photosynthesis is the process by which plants convert carbon dioxide into their food by using the energy derived from the sun. The most essential elements of this process are sunlight, water, carbon dioxide, and chlorophyll. While carbon dioxide is absorbed by leaves, water enters the plant through its roots. After being absorbed by the roots, water travels through the stem to reach the leaves where the actual process takes place. Carbon dioxide and water react in the presence of light energy to produce glucose and oxygen, which are burned (respiration), producing carbon dioxide.

always had—no more, no less, just transiently stored in different places.

Like the earth's water cycle (evaporation—condensation—precipitation), there is a carbon cycle, a methodical series of natural events that allow for carbon atoms to be reused as they pass from Earth to the atmosphere and back again, over and over. Most carbon is stored in rocks and sediments, while the rest is reserved in the ocean, atmosphere, and living organisms. These are the reservoirs, or *sinks*, through which carbon constantly cycles. For example, the ocean (71 percent of Earth's surface) is a colossal sink that actively absorbs excess atmospheric carbon. The carbon from dead and decaying organic marine organisms and creatures also settles in the waters. Over the ages, carbon stored in the many sinks wonderfully develops into fossil fuels. When humans utilize carbon-based fossil fuels for energy, CO_2 is released back into the atmosphere and continues to be recycled.

Working in concert with the other atmospheric greenhouse gases, carbon dioxide assists in keeping our planet habitable by absorbing the sun's immense heat and regulating ambient air temperature. Without this vital shield held in our skies, we would experience wild diurnal temperature swings varying one hundred degrees between night and day.

This, however, is where climate activists will stop the conversation, insisting that the use of fossil fuels dangerously exacerbates the greenhouse effect, trapping an overabundance of heat within the atmosphere and artificially warming the planet, throwing the climate out of whack.

The proverbial doomsday clock is ticking, they insist. That's why, they contend, we must "follow the science."

REAL SCIENCE VS. CONSENSUS

But that phrase, "follow the science," is intentionally misleading.

A friend, Dr. Dave, a professor of neurobiology, told me, "*Follow the science* implies some type of monolithic *thing.*" Explaining further, Dr. Dave said, "Science is really at its best as a word when used as a verb. We *do* science. And an act of doing science is only as good as two essential elements: first, the logical quality of the reasoning involved and, second, the quality of the premises as evidence and data being input at the front end of that process. If one is not responsible with this process, it's garbage in, garbage out.

"*Follow the science,*" he continued, "has become code for *accept the groupthink.*"

That's *not* how science works.

Science is a rigorous process that follows a routine method meant to be scrutinized. It begins with a hypothesis: a suggested explanation for an observable phenomenon or a reasoned prediction of a possible causal correlation among multiple phenomena.

In the case of anthropogenic global warming, or climate change, the hypothesis would state:

> *If* fossil fuels (which emit carbon dioxide into the atmosphere when consumed) are utilized for the production of energy, *then* such usage will artificially warm the atmosphere, altering the earth's climate and endangering life on earth, *because* carbon dioxide is a greenhouse gas.

Note, the hypothesis contains an *if*, *then*, and *because*.

Next, the hypothesis would be tested properly and repeatedly through observation, data collection, analyzation, and interpretation in an attempt to disprove the hypothesis. A quote attributed to Albert Einstein explains it this way: "No amount of experimentation can ever prove me right; a single experiment can prove me wrong."

If the hypothesis withstands the tests, the results are published for others to see so they may conduct further study to disprove it. At this point, the hypothesis could progress to a theory, which is a further-tested, well-substantiated, unifying explanation for a set of verified, proven factors. But even that theory would continue to be vigorously investigated. If eventually proven rock solid, a scientific theory can become law, as in the laws of gravity, motion, or thermodynamics.

Anthropogenic (human-caused) climate change advocates contend their theory has reached the crucial state of absolute verification, and therefore profound action must be taken immediately to save the human race.

Cooler minds argue, "Not so fast."

People often say "theory" when they are actually referring to a hypothesis. For instance, someone could claim they have a theory as to why they have trouble sleeping: "If I drink coffee after two in the afternoon, then I have problems sleeping at night, because caffeine in the coffee is a stimulant." The statement is based on simple observation and instantly registers with a lot of people, but sleeping problems and insomnia can be caused by a multitude of factors like stress, medications, alcohol, spicy foods, electronic devices, neurological problems. Since this observation is a reasonable possibility, it is testable and can be falsified, which makes it a hypothesis, not a theory.

So, returning to the climate change hypothesis, carbon dioxide emissions from fossil fuels are indeed a greenhouse gas. But I am about to share something the cult leaders do not want you to hear: CO_2 is a gas that has historically varied in concentration and only represents a *sliver* of the overall atmosphere, and contribution to this sliver by humankind's use of fossil fuels is miniscule. Additionally, there are other potent greenhouse gases that need to be considered; and there is a vast carbon sink—the oceans—that absorb CO_2. Plus, over time, there are always major ebbs and flows in solar radiation, temperature, and climate. Also, if the cause of global warming actually is the emission of CO_2 into the atmosphere by way of fossil fuels, a rise in CO_2 should precede the rise in temperature. But that is not what is observed in the historical record. It must also be noted that there are many sources of CO_2 other than our use of fossil fuels—volcanic eruptions, forest fires, smoldering peat bogs, animal (and human) breathing, and decay of deceased plant and animal matter.

But simply even pointing those facts out quickly earns one the label of *denier*.

In an interview on *The Joe Rogan Experience* podcast, Dr. Steven Koonin, former under secretary for science at the US Department of Energy during the Obama administration and author of *Unsettled: What Climate Science Tells Us, What It Doesn't, and Why It Matters*, said that he has met quite a few colleagues who are privately skeptical of the anthropogenic climate change theory, but they refuse to allow their doubts be made public, obviously because they fear the personal and professional consequences.[6]

---

6   *The Joe Rogan Experience*, episode 1776, interview with Steven E. Koonin, February 14, 2022.

You see, many climate scientists are aware that *garbage in, garbage out* rakes in funding.

The global warming/climate change assumption should remain subject to honest testing because it has in *no way* ascended to scientific law. Yet, the politicians, policymakers, influencers, and activists accept inaccuracy to offer up profound plans to "save" the world, programs that will reestablish society as we know it. It's like someone with an exceptionally low risk of dying prematurely going into debt to purchase a crème de la crème long-term health care insurance policy with outrageous monthly premiums.

To be sure, atmospheric carbon dioxide levels have increased since 1800 (though well within the geological record). And yes, the temperature has slightly warmed since something known as the Little Ice Age ended about 1750. But there are so many people of esteem in their fields who remain quite skeptical of the absolute application of the climate change contention. For example:

- The late Dr. Fred Singer, former space scientist and meteorologist, foundational in the development of weather satellites.

- Dr. Neil Frank, former director of the National Hurricane Center.

- Patrick Moore, co-founder of Greenpeace.

- Dr. John Clauser, 2022 Nobel Prize-winning physicist.

- Dr. Joel Myers, meteorologist, founder of AccuWeather.

- John Coleman, founder of the Weather Channel.

- Dr. Roy Spencer, principal research scientist for the University of Alabama in Huntsville, known for overseeing satellite-based temperature monitoring.

- The late Dr. Bill Gray, hurricane research pioneer and forecaster.

- Dr. Joanne Simpson, one of the most preeminent scientists of the last one hundred years and the first woman to obtain a PhD in meteorology.

- Dr. Harrison Schmitt, a Harvard-trained geologist and one of twelve people who have walked on the moon.

- Willie Soon, astrophysicist, Harvard-Smithsonian Center for Astrophysics.

- Richard Lindzen, atmospheric physicist, professor, Massachusetts Institute of Technology (retired).

Those are just a few, and there are hundreds more. But those leading the masses into the cult would tell you these people are deniers, outliers, and a distinct minority, as if to imply scientific theory is brought about by consensus.

History proves it does not work like that.

COPERNICUS

Around 150 AD, Claudius Ptolemy, an Egyptian living in Alexandria, gathered and organized the thoughts of earlier scientific thinkers and published his theory of the universe. Ptolemy believed the earth was a fixed, immovable mass, located at the very center of the universe, and all celestial bodies, including the sun and the stars, revolved around it.

Since the telescope would not be invented for another 1,500 years, it was difficult in Ptolemy's time to conclusively dismantle his theory, which eventually became known as the Ptolemaic system and was taught in all institutions of advanced learning as scientific law. Questioning Ptolemy's view of the universe carried severe consequences including excommunication from the Roman Church, banishment from one's country, and, depending on the "heretic's" temperament, ecclesiastic authorities could prescribe a public burning at the stake.

Because of the intense pressure to conform to Ptolemy's worldview, no debate was tolerated, thus, giving an appearance of consensus. Skeptics were forced to secretly discuss their own contradictory theories of the universe in the shadows, as religion and science were so integrated in that society.

Finally, 1,400 years later, there was a major breakthrough.

Polish mathematician and Roman Church official, Nicolaus Copernicus, spent decades privately defying the centuries-old cover-up. Through extensive research, much of which was conducted under the cover of night atop the roof of his church and without the aid of a telescope (that invention was still sixty years away), Copernicus asserted that the earth rotated on its axis once daily and traveled around the sun once annually.

Planet Earth was not the hub of the universe.

Copernicus's findings were quietly presented in 1530 in his brilliant research manuscript, *De Revolutionibus Orbium Coelestium* (*On the Revolutions of the Heavenly Spheres*), which he privately shared with trusted colleagues who were intrigued and challenged by his detailed work. Fortunately, Copernicus died naturally in 1543. Had he lived another decade or two, his last breath may well have been taken from him by the flames of the stake.

Following his passing, *De Revolutionibus* began to have an impact on the thinking of many subsequent great minds. Italian scientist Giordano Bruno was convinced of the Copernicus system and, like an inspired evangelist, dared to openly share the theory. For his zeal, Bruno was arrested by religious authorities in 1592 and placed on trial. Refusing to renounce his beliefs in Copernicus's principles, Bruno was burned at the stake, and *De Revolutionibus* was placed on the list of forbidden books.

Galileo Galilei was another Ptolemy doubter. Utilizing observations gathered with his new invention, the telescope, in 1632, he boldly published, *Dialogue Concerning the Two Chief World Systems*, a work that upheld the Copernican system over Ptolemy's. Galileo was brought before the doctrinal inquisitors in Rome who commanded him to renounce all beliefs and writings supporting Copernicus's theory—but it was too late. Though very primitive, Galileo's telescopes were quickly being mass-produced, and too many eyeballs were confirming the truth that our planet and solar system were plainly rotating about the sun. The mandated consensus, which had held fast for nearly 1,500 years, was finally broken.

A brilliant definition of consensus was once given by former British prime minister Margaret Thatcher:

> "The process of abandoning all beliefs, principles, values, and policies in search of something in which no one believes, but to which no one

objects; the process of avoiding the very issues that have to be solved, merely because you cannot get agreement on the way ahead. What great cause would have been fought and won under the banner: 'I stand for consensus'?"[7]

Science is not accomplished by consensus.

CO_2 AND THE PLANT KINGDOM

The average person only hears one side of the climate story, the side that says CO_2 from cars, trucks, gas stoves, gas heaters, and industry at large is a *dangerous* greenhouse compound that is causing global warming and climate change. They will persuasively call your attention to news reports of heat waves, storms, and droughts as routine proof while trying to turn your imagination to the invisible culprit: carbon dioxide.

But some compelling facts reveal a different story. Remember, plant life absorbs CO_2 and emits oxygen. So, how would the plant kingdom respond to an *increase* in carbon dioxide?

Research conducted by Michigan State University professor emeritus of horticulture, Sylvan H. Wittwer,[8] indicates plants thrive in an atmosphere with raised levels of the gas. For example, with a tripling of CO_2, roses, carnations, and chrysanthemums experience earlier maturity, have longer stems, and larger, lon-

7   Margaret Thatcher, quoted in Ethan Siegel's article, "What Does 'Scientific Consensus' Mean?" *Forbes*, June 24, 2016, https://www.forbes.com/sites/startswithabang/2016/06/24/what-does-scientific-consensus-mean/?sh=179b9edf6bae.

8   Sylvan H. Wittwer, "Rising Carbon Dioxide Is Great for Plants," *Policy Review*, Fall 1992.

ger-lasting, more colorful flowers with yields increasing up to 15 percent. Rice, wheat, barley, oats, and rye perform yield increases ranging to 64 percent. Potatoes and sweet potatoes produce as much as 75 percent more. Legumes, including peas, beans, and soybeans, show increases to 46 percent.

The effects of carbon dioxide on trees, which cover one-third of the earth's landmass, may be even more dramatic. According to Michigan State's forestry department, trees have been raised to maturity in months instead of years when the seedlings were raised in a tripled CO_2 environment.[9]

This is known as CO_2 enrichment or fertilization. It can be accomplished with the release of compressed carbon dioxide or with a CO_2 generator. With the growth of the greenhouse industry and indoor gardening since the 1970s, this process is frequently used to increase plant growth and yields. Even commercial marijuana growers utilize this method.

Any increase in atmospheric CO_2 would be well-received by the plant kingdom. And if carbon dioxide were truly warming the climate, wouldn't such change expand the earth's temperate zones, allowing for more precious plant life to grow and flourish?

Step back and ask yourself, "How could something so essential to life be so vilified?"

These are important questions. And here's another: How could a gas like CO_2, which is so miniscule in comparison to all the other gases in our atmosphere, be such a threat?

---

9   Ibid.

SO LITTLE CO_2

Critics will foment over this section, because, like slick salesmen selling an imperfect product, they do not want you to take a good look at what you are buying. It's astounding to note that of all the gases in our atmosphere, the amount of CO_2 is almost imperceptible. By percentage, the gases are ordered as follows:

Nitrogen	78.1 percent
Oxygen	20.9 percent

Note: nitrogen and oxygen comprise 99 percent of all atmospheric gases.

Water vapor	0.40 percent[10]
Argon	0.9 percent
Carbon dioxide	0.04 percent
Neon	0.002 percent
Helium	0.0005 percent
Methane	0.0002 percent
Krypton	0.0001 percent
Hydrogen	0.00005 percent
Nitrous oxide	0.00003 percent
Ozone	0.000004 percent
Carbon monoxide	a minute trace

Carbon dioxide only accounts for *four hundredths of a percent* of our planet's atmosphere.[11] It is also known as a variable gas

---

10   The total atmospheric amount of water vapor always varies greatly.
11   Often noted as 400 ppm, or 400 parts per million.

(along with water vapor) because, as mentioned, historically it fluctuates. And what percentage of the infinitesimal amount of CO_2 is produced by human activities including the utilization of fossil fuels? The answer to this vital question is one the climate activists do not want you to comprehend.

When I wrote *Climategate* in 2010, the CO_2 percentage was readily available online via any search engine. An analysis by the Carbon Dioxide Information Analysis Center (at the time a research wing of the US Department of Energy) reported that human contribution of atmospheric CO_2 was only 3.207 percent,[12] easily deflating the climate agenda's sails. However, the analysis center was shuttered in 2017, and since then, such data has been virtually scrubbed from the internet; now you have to figure the math yourself. So, using high estimates, human activity may now contribute to something in the neighborhood of 4 percent of the four hundredths of a percent. In other words, of all the gases in the atmosphere, 0.0016 are placed there by human activities. As a Silicon Valley mathematician explained to me, for every sixty-two thousand molecules in the sky, only one is there because of humankind's use of fossil fuels.

If that's too small to wrap your brain around, then let's try another analogy.

On page 387 of the hardcover edition of Michael Crichton's bestselling eco-thriller, *State of Fear*, Crichton likens the gases of the earth's atmosphere to a football field. It is a great word picture that is belittled by climate activists.

---

12 "Current Greenhouse Gas Concentrations," updated October 2000, Carbon Dioxide Information Analysis Center, US Department of Energy; this department was later shut down in 2017.

From the goal line to the 78-yard line contains nothing but nitrogen; oxygen fills the next 21 yards, stretching to the 99-yard line. The final yard, except for four inches, is argon, a wonderfully mysterious inert gas useful for putting out electronic fires. Three of the remaining four inches are crammed with a variety of minor, but essential, gases.

And the last 1.44 inches?

Yep. Carbon dioxide. Less than an inch-and-a-half of a hundred-yard field! If you were in the stands looking down at the turf, you would need high-powered binoculars to see the width of that line.

And the most crucial point: How much of that last inch is contributed by human activities?

The equivalent of a line thinner than a dime standing on edge.[13]

So, how much has CO_2 increased in the atmosphere since the use of fossil fuels began in the early nineteenth century?

Approximately 50 percent—or half the thickness of a dime.

During over two hundred years of industrialization, with the population of the planet increasing from approximately one billion to eight billion, with billions of cars, trucks, airplanes, and ships, all with engines running on fossil fuels, and billions of homes and buildings powered by electricity derived in large measure from natural gas and coal, the "pollutive gas" CO_2 remains a sliver.

And, the increase in CO_2 is clearly within historical norms. This is important to note because the climate change influencers

---

13   In *State of Fear*, Crichton uses a quarter in the analogy, but the actual thickness attributed to the human contribution in this demonstration would be 1.015 mm, so the better coin choice would be a dime, which is 1.35 mm.

are banking on our collective shortsightedness-based observations made during our brief human lifespans. Profound atmospheric changes occur not just on a decadal basis but over centuries, millennia, and beyond.

Paleoclimate researchers are quick to reveal data illustrating that in the Eocene period (fifty million years ago), CO_2 was likely six times higher than today. In the Cretaceous period (ninety million years ago), it was as much as seven times higher than today, and in the appropriately named Carboniferous period (340 million years ago), carbon dioxide was thought to be nearly twelve times higher than current levels.[14] Many surmise dinosaurs in the Jurassic period were able to maintain such behemoth sizes because of the indescribable abundance of foliage fed by the heightened levels of CO_2 present during that era. For most of the earth's history, the concentration of CO_2 has been significantly higher than today. And by the way, the historical record does not bear out extinction events by increased levels of CO_2.

GREENHOUSE GAS GAME

The "greenhouse effect" is aptly named. First coined by scientists in the 1800s, it describes the way crucial gases in our atmosphere absorb heat from the sun, thus maintaining an environment appropriate for human habitation. Scientists in the nineteenth century used the term in a favorable way, rightly conjuring relatable imagery of the warmth experienced in a flower or vegetable greenhouse. With its glass walls and roof, a greenhouse allows the sun's heating rays to shine through and enter the otherwise shel-

---

14   C. J. Yapp and H. Poths, "Ancient Atmospheric CO_2 Pressures Inferred from Natural Goethites," *Nature*, vol. 355, no. 23 (January 1992): 342–344.

tered environment, warming it nicely in comparison to conditions outside the glass. In addition, the greenhouse traps an entire day's worth of warmth, preventing the heated air from completely radiating back into the sky at night. When soil is brought into the greenhouse, seeds sown, and irrigation applied, other factors begin to dramatically warm the artificial environment. The water from irrigation begins to evaporate, creating vapor and increasing the humidity to further warm the surroundings.

Depending on where you reside, you may be well acquainted with the effects of humidity. On a humid summer day, it's difficult to move about outdoors without beads of perspiration forming on your brow. At night, the water vapor-laden air seems heavy, and the temperature has a difficult time dropping to comfortable sleeping levels. This is because humid air tends to retain its temperature; it's the greenhouse effect working at the microclimate level—in this case, where you live. On the macroclimate level, without the greenhouse effect, the earth would be a desert devoid of life.

Of the atmosphere's gases, only five are greenhouse gases. Not surprisingly, many alarming websites touting climate change fail to list the herculean of these five in their frightful commentaries: water vapor. This is incredibly disingenuous given that water vapor accounts for 95 percent of those greenhouse gases. Perhaps water vapor is not mentioned because the computer models used to predict the long-term consequences of anthropogenic warming are incapable of representing the complexities presented by condensed water vapor in the form of clouds. As Dr. Roy Spencer, one of the eminent scientists noted above, writes:

> One of the primary methods the Earth [has] for cooling itself is the production of clouds, which reflects some of the solar radiation that reaches the Earth back to outer space. Because the average effect of clouds on the Earth's surface is to cool it, any natural change in global average cloudiness can also be expected to cause global warming or global cooling.[15]

After water vapor, the remaining 5 percent of greenhouse gases are, in order, CO_2, methane, nitrous oxide, ozone, and carbon monoxide. However, it must be noted that methane is 21 times more potent than CO_2 when it comes to the greenhouse effect, and nitrous oxide is 310 times more capable of retaining the sun's heat than CO_2.

Yet the climate change hypothesis implies the portion of CO_2 from human activity (4 percent) somehow controls the rest of CO_2 (96 percent). The hypothesis also contends CO_2 to be the main controller of temperature and climate.

This is unfounded.

Carbon dioxide is actually a lesser player in the greenhouse game, but it represents something that political and cultural progressives seem to loathe, *happiness*—happiness that comes from our cars and trucks, the heating and air conditioning units used to keep our homes comfortable, the tractors plowing fields for our bountiful food supply, the airplanes we fly in to see loved ones and take vacations, the industries that manufacture the products we enjoy. It's von Liebig's recycled guano argument all over again,

---

15 Dr. Roy W. Spencer, *The Great Global Warming Blunder: How Mother Nature Fooled the World's Top Scientists* (Encounter Books, 2010), 21.

proclaiming from the grave that those of us who desire a better quality of life are guilty of maintaining "a robbery system."

RECYCLING CO$_2$

When a major volcano blows its lid on the Pacific Rim, a lightning-induced forest fire rages in the Rockies, or an Indonesian peat bog continually smolders, they naturally release huge amounts of long-stored CO_2 into the atmosphere. The carbon dioxide banked in weathering rocks, decaying coral, and decomposing plants is also constantly meandering through the cycle.

When the carbon cached in fossil fuels is released back into the atmosphere, it's temporarily held and finally absorbed by a variety of sinks, like the ocean. Dissolved carbon dioxide, then, becomes the exoskeleton of the snails, shellfish, and coral. The ocean floor is also rich in sedimentary limestone—a "petrified" modification of CO_2, also known as calcium carbonate (the same stuff you take to ease a stomachache or heartburn).

Another major sink includes the organic carbon compounds found in all things both alive and dead. Deceased organic matter includes coal seams, natural gas, and petroleum reserves as well as newly fallen autumn leaves, recently felled trees, and even animal and human corpses. Through the process of decay, the carbon stored in these substances is released back into the air as inorganic carbon dioxide to be reworked into the carbon cycle.

There is also a mystery at play in the carbon cycle that scientists have a tough time computing. When an agitation occurs within the carbon cycle—for example, a major volcanic eruption—natural mechanisms seem to maintain the cycle's equilibrium. This was noted when Mount Pinatubo erupted in the

Philippines in 1991. The enormous ash plume from that volcano was visible on satellite imagery and shown by every television meteorologist in America. A cloud of particulate eventually covered much of the earth. Over the next two years, the earth's average temperature dropped by one-half degree Fahrenheit.[16] The particulate matter injected into the atmosphere by Pinatubo was able to partially block the sun's radiation and, thus, decrease the global temperature. But the mysterious conundrum for scientists is that during the same two-year period, amounts of atmospheric carbon dioxide *decreased* globally.[17]

Could it be that the massive amounts of CO_2 spewed into the sky by one of the most powerful volcanoes in our lifetime was being offset by some natural mechanism like absorption by the oceans?

This is why we test hypotheses and theories.

CONCLUSION

In 2012, a team from the PBS documentary program, *Frontline*, attended a climate conference hosted by the Heartland Institute. I was honored to be a speaker at the event, sharing the podium with several of the experts listed earlier in this chapter. A PBS producer really wanted to get an interview with Dr. Fred Singer. Dr. Singer was hesitant. But since I was the guy with extensive media experience, he consulted with me before agreeing to go before the

16   NASA/Goddard Space Flight Center, "Mt. Pinatubo Eruption Provides A Natural Test For The Influence Of Arctic Circulation On Climate," ScienceDaily, March 13, 2003, https://www.sciencedaily.com/releases/2003/03/030313081900.htm.

17   News Release, "Large Volcanic Eruptions Help Plants Absorb More Carbon Dioxide from the Atmosphere," NASA/Goddard Space Flight Center, December 10, 2001, https://www.eurekalert.org/news-releases/872996.

camera. We spoke with the PBS crew for a few minutes and then Dr. Singer accepted the interview, portions of which were used in a program titled "Climate of Doubt."[18]

Some quotes that were left on the cutting room floor:

> "Climate change is a natural phenomenon. Climate keeps changing all the time. The fact that climate changes is not in itself a threat, because, obviously, in the past, human beings have adapted to all kinds of climate changes.

> "…as carbon dioxide increases, you would expect a warming. But at the same time that you get this warming or this slight warming, you get more evaporation from the ocean. That's inevitable. veryone agrees with that. Now, what is the effect of this additional water vapor in the atmosphere? Will it enhance the warming, as the models now calculate? Or will it create clouds, which will reflect solar radiation and reduce the warming? Or will it do something else? You see, the clouds are not captured by the models. Models are not good enough to either depict clouds or to even discuss the creation of clouds in a proper way. So, it's not possible at this time to be sure how much warming one will get from an increase in carbon dioxide.

> "…I think the warming will be much less than the current models predict. Much less. And I

---

18   PBS, *Frontline*, Season 2012, Episode 20, aired October 22, 2012.

think it will be barely detectable. Perhaps it will be detectable, perhaps not. And it certainly will not be consequential. That is, it won't make any difference to people."[19]

There are many other researchers, particularly outside of the United States, who contest the theory of anthropogenic climate change and unabashedly challenge their peers on the other side of the debate. For example, here is the summary statement from an excellent recent study, conducted by professional engineers from Great Britain and Germany, published in the *International Journal of Atmospheric and Oceanic Sciences*:

> ...the increasing levels of CO_2 will not lead to significant changes in earth temperature.... This is an unfortunate situation since world governments are implementing ambitious and expensive plans to limit the increase of global temperatures by reduction of CO_2 emissions to atmosphere, and to achieve a net zero carbon economy, in the belief that CO_2 emissions are the main driver of global temperature increases....[20]

The research team's conclusion parallels another study published a year earlier, "Comprehensive Analytical Study of the Greenhouse Effect of the Atmosphere," by Dr. Peter Stallinga of

---

19   PBS, *Frontline*, Interview with Dr. S. Fred Singer, https://www.pbs.org/wgbh/warming/debate/singer.html.

20   David Coe, Walter Fabinski, Gerhard Wiegleb, "The Impact of CO_2, H_2O and Other 'Greenhouse Gases' On Equilibrium Earth Temperatures," *International Journal of Atmospheric and Oceanic Sciences*, vol. 5, no. 2 (2021): 29–40.

Portugal[21] who states, "...there is nothing CO_2 would add to the current heat balance in the atmosphere."[22]

At the very least, such conclusions should cause a legitimate truth-seeker to pause and reconsider the most popular scientific assumption of our lifetime. It has gotten to the point where CO_2 "pollution" now drives an entire movement dedicated to completely resetting society as we know it.

---

21 Dr. Stallinga is a professor of science and technology at Universidade do Algarve, Faro, Portugal. He did his postdoctoral studies in materials science and engineering at University California, Berkeley and the Department of Physics and Astronomy, Aarhus University, Denmark.

22 Dr. Peter Stallinga, "Comprehensive Analytical Study of the Greenhouse Effect of the Atmosphere," *Atmospheric and Climate Sciences*, vol. 10, no. 1 (January 2020): 40–80.

CHAPTER FOUR

UNTIL NOW

WHEN I WAS A TV meteorologist, I loved doing weathercasts outdoors. Outdoors is where the weather happens, right?

"You're doing a live remote tomorrow, Brian," the executive news manager told me. "It's going to be a hundred and five."

"Let me guess, the bank in Walnut Creek," I replied sarcastically.

I'd been through this drill so many times. Walnut Creek was the perfect location. We had lots of viewers in that area. It is an upscale town thirty miles east of San Francisco, sheltered from the cooling influences of the Pacific Ocean and the San Francisco Bay by a modest mountain range. As a result, summer temperatures can soar. The bank not only referenced the name of the town but had a thermometer that always produced a readout several degrees higher than accurate, due to the bank's paved parking lot with its heat-absorbing asphalt as well as that of the adjacent multilane boulevard; 105°F would easily be seen as 110°F.

Whenever I did this routine in front of the bank, I would always explain the exaggerated reading for our audience. But even with the explanation, the visual was tough to overcome. Remember, tele-*vision*.

Viewers without fail would comment, "Did you see that? One-ten in Walnut Creek!"

Everyone at the TV station knew extreme weather's always a bestseller, especially with the increased hype over global warming. So, the deceptive bank thermometer always bothered me, the "uptight" weather guy. But TV news thrives on visuals.

TAKING THE TEMP

But here's the problem: it's not just faulty bank thermometers. The earth's entire temperature record can be misleading. Most people naïvely assume the atmosphere is accurately measured by trustworthy scientists who are concerned about precision and credibility—but it's way more complicated than that. And this is where researcher bias can subtly skew results, resulting in some deceptive conclusions.

Earth's temperature record is relatively recent. Besides the telescope, Galileo invented a rudimentary water-based thermometer in 1593, which, for the first time, allowed temperature variations to be accurately measured. In 1714, Gabriel Fahrenheit devised the first mercury thermometer—the same technology commonly used to measure air temperature today. By the 1800s, mercury thermometers were being mass-produced and used to record the highs and lows around the world.

So, for starters, we've only been accurately keeping tabs on regional temperatures in various parts of the world for a few hundred years. But there are other glaring inconsistencies. For one, thermometer coverage has always been distressingly sparse. And because of political upheaval and the ravages of war, weather logs that do exist are filled with glaring gaps of missing and unreliable data.

That said, the most consistent long-term monitoring has been conducted in the United States. It is known as the US Historical Climatology Network (USHCN), and it includes 1,218 government-sanctioned monitoring stations possessing uniform equipment and guidelines. Most site logs contain entries dating back to the 1800s.

The USHCN instruments are individually and manually maintained by perpetual stewards: farmers, ranchers, firehouse staff, municipal workers, students, and park rangers. There is no automation with this network; each thermometer is visually observed twice daily to accurately determine the maximum and minimum readings (the thermometers possess a clever device that marks the high and low reading, thus, eliminating any guesswork).

If you were to average all the data from the 1,218 locations, you would note a mere increase in temperature of about one-half degree Fahrenheit since 1900. This is noteworthy, because an average of all the other measurements supposedly indicates we have experienced an increase of 1.8°F (1°C) since the turn of the twentieth century. Some scientists insist the rise is even higher (1.2°C).

However, even within the USHCN archives there are serious problems suggesting the one-half degree increase is inflated. For example, occasionally life happens, and observers fail to observe. A few lazy stewards were discovered to have placed light bulbs in the thermometer shelters in order to log the afternoon's maximum temperature after sunset—forget to flip the light switch off, and they've spoiled the next day's readings.

But a more widespread problem has been something known as "the urban heat island." Case in point: Since 1835, USHCN readings have been regularly logged in New York City's Central Park. In 1835, the city's population was only a bit more than two hundred thousand. Today, however, Central Park is surrounded by a metropolis filled with some of the world's tallest buildings, a massive underground subway, extensive sewer systems, power generation facilities, and millions of cars, buses, taxis, and Ubers transporting eight million residents and forty-five million tourists each year on asphalt streets. That infrastructure absorbs a tremendous amount of heat and skews maximum and minimum temperature readings. Thus, Central Park's historical temperature graph illustrates an incredible warming increase of nearly 4°F since the 1800s.

When examining an equally mature USHCN temperature record just fifty-five miles away at the West Point Military Academy, where the grounds have changed minimally by comparison, the difference is glaring—West Point has only warmed one-half degree Fahrenheit over the same time period.

And then there is a USHCN station charged with monitoring the beautiful Lake Tahoe, California, basin. Records there indicate that since the early part of the twentieth century, temperatures have risen some 4°F. But the data is corrupted.

Anthony Watts has been doing a yeoman's job in determining how many of the Historical Network's weather sites have been ruined. His premier website, WattsUpWithThat.com, has a section dedicated to sleuthing inaccurate monitoring stations around the world.

Watts discovered that for many years, the Tahoe site used a fifty-five-gallon drum for incinerating trash that was located five feet away from the thermometer shelter. Adjacent the incinerator was also an assortment of heat-absorbing junk metal. In the eighties, a tennis court was installed some twenty-five feet away (by regulation, weather stations are to be one hundred feet from any paved surface), and in the summer, the station is surrounded by parked vehicles.

Any wonder how the Tahoe readings always showed a hot trend?

In addition to the Historical Network, there exists a system of over one thousand automated weather sites known as the Automated Surface Observation System (ASOS), mostly installed at airports. These sites are funded and maintained by the federal government in conjunction with the National Oceanic and Atmospheric Administration (NOAA), National Weather Service, Federal Aviation Administration, and the Department of Defense. It is a newer system with most of its monitors installed in the 1980s. Like the bank thermometer, though, much of ASOS's information has been compromised by the urban heat island.

In addition to the data collected in the United States, temperatures are culled globally from a thin network of ocean-based weather stations, plus a collection of unequally distributed, ragtag readings scattered around the world—with the largest percentage having been installed since the 1980s. The quality control on this world network is far from optimum.

A third method of taking the earth's temperature—certainly the most effective one, as it can cover all portions of the earth—is with satellites (more specifically, satellite-based microwave radiation measurements). It's a technology that began in 1979 with the previously mentioned Dr. Spencer from the University of Alabama in Huntsville (UAH) at the helm. But even this method is not perfect, as readouts must be continually adjusted to account for deficiencies within the technology.

While Spencer and his colleague John Christy have received numerous accolades for their work, they have also attracted controversy because their satellite record never revealed an alarming warming trend. That changed about twenty years ago when scientists at Remote Sensing Systems (RSS) in California discovered a glitch in the UAH algorithm that, once revised, yielded the warming trend.[1]

Spencer and Christy incorporated the RSS correction, but the two teams subsequently differed on other questions, like how to correct for the constant positional drift of the satellites, which alters the time of day when instruments take their readings over each location. Another concern was how to merge records when one satellite is taken out of service and replaced by a new one (incorrect splicing of data compilation can introduce spurious warming or cooling).

Nonetheless, RSS has consistently exhibited more warming than UAH.

Until recently.

---

1   Professor Ross McKitrick, "The Important Climate Study You Didn't Hear About," *Fraser Institute*, April 12, 2023, https://www.fraserinstitute.org/article/the-important-climate-study-you-didnt-hear-about.

HISTORICAL FLUCTUATIONS

Karl Marx said, "History does nothing; it possesses no immense wealth; it wages no battles."[2]

The progressive planners pushing the climate agenda do not want their faithful flock to realize that so many tragic zeniths of human history repeat themselves. And, since members of the world's elite are employing climate as the cornerstone of their scheme to reset society, like Marx, they certainly do not want those with "lesser minds" to be aware of Earth's environmental history.

Long before fossil fuels and carbon dioxide emissions were present, there have been deadly periods of drought. According to the *Glossary of Meteorology* (1959), a drought is defined as "a period of abnormally dry weather sufficiently prolonged for the lack of water to cause serious hydrologic imbalance in the affected area." Droughts impact water supplies, cause crops to fail, and can lead to mass starvation. Droughts have altered history. The following are five of the world's worst droughts, none of which can be blamed on anthropogenic climate change:

- On the African continent 75,000 to 135,000 years ago, there were "megadroughts." Anthropologists suspect these may have been responsible for the first human migrations from the continent.

- Archaeologists investigating the royal tombs of Egypt's Old Kingdom discovered evidence of a severe drought that struck the Middle East and portions of Europe 4,500 years ago. Some experts conclude it was that drought,

---

2   Karl Marx, *The Holy Family*, Chapter 6 (1845).

rather than civil upheaval, that caused the fall of the pharaohs, who ruled ancient Egypt for three thousand years before the region became a province of the Roman Empire in 30 BC.

- A massive drought hit the Mayan empire over 1,200 years ago in Mesoamerica (present day southern Mexico, Guatemala, Honduras, Belize, El Salvador, and Nicaragua). Dwindling water resources caused a war with neighboring nations that brought the demise of the Mayan civilization.

- China's most disastrous drought occurred in 1928–1930, causing widespread famine and claiming the lives of perhaps up to ten million people.

- The North American Dust Bowl in the Great Plains of the Midwest and Canada in the 1930s drove two million people from their homes and led to an outbreak of deadly disease. Many record temperatures established during that decade still stand today.

- Droughts happen. Naturally. And so does hot weather.

HOTTEST DECADE EVER

"You can run, but you can't hide," a phrase that has ended up in many songs and movies, but in reality, it's a paraphrase of a comment made by former heavyweight boxing champion Joe Louis, who held the title from 1937–1949.

In 1941, prior to a victory over Billy Conn, a heavyweight known for fancy footwork, Louis told the press, "He can run, but he can't hide."

And, so it goes with the statistics you are about to read. Climate change advocates can't hide from these statistics, so they, instead, run with spurious explanations that always manage to place the blame squarely on mankind.

One of the glaring historical examples they'd like to hide from is the North American Dust Bowl, which recorded the hottest decade ever in the US.

The summer of 1930 marked the beginning of the longest drought of the twentieth century, and it began with a bang. From June 1 to August 31, 1930, Washington, DC, experienced twenty-one days with maximum temperatures reaching at least 100°F. Several of those record readings remain unbroken nearly a century later.[3]

By 1934, bone-dry regions stretched from New York, across the Great Plains, and into the Southwest. The following year, a miserable "dust bowl" covered about fifty million acres in the south-central plains. Five-thousand people would perish due to the inclement weather.

And it only got hotter.

According to the National Climatic Data Center, 1936 experienced the hottest overall summer ever recorded in the continental United States. All-time records for many states, and the District of Columbia, that were established in the 1930s remain standing today:

---

3   In 1945, the location of the original weather station in Washington, DC, was moved to the municipal airport.

*110°F Millsboro, Delaware, July 21, 1930

*106°F Washington, DC, July 20, 1930

*114°F Greensburg, Kentucky, July 28, 1930

*113°F Perryville, Tennessee, August 9, 1930

*100°F Pahala, Hawaii, April 27, 1931

*109°F Monticello, Florida, June 29, 1931

*118°F Keokuk, Iowa, July 20, 1934

*116°F Moorhead, Minnesota, July 6, 1936

*121°F Steele, North Dakota, July 6, 1936

*109°F Cumberland and Frederick, Maryland, July 10, 1936

*111°F Phoenixville, Pennsylvania, July 10, 1936

*110°F Runyon, New Jersey, July 10, 1936

*112°F Mio, Michigan, July 13, 1936

*114°F Wisconsin Dells, Wisconsin, July 13, 1936

*116°F Collegeville, Indiana, July 14, 1936

*112°F Martinsburg, West Virginia, July 19, 1936

*121°F Alton, Kansas, July 24, 1936

*118°F Minden, Nebraska, July 24, 1936

*120°F Ozark, Arkansas, August 10,1936

*114°F Plain Dealing, Louisiana, August 10, 1936

*120°F Seymour, Texas, August 12, 1936

*117°F Medicine Lake, Montana, July 5, 1937.

And the blistering heat of the thirties lingered into the 1940s.

While it is clearly foolish to blame carbon dioxide emissions for that scorching weather, faithful climate change disciples have been taught to blame humans anyway. A high school resource article produced by *National Geographic* quotes a professor of earth science at the University of Pennsylvania who contends, "...[T]he human fingerprint on the Dust Bowl was agricultural practices and our ignorance of the nature of the Great Plains."[4]

While that excuse might work to explain the dust, it hardly does so for the heat.

Quite suddenly, however, the thirties' hot trend reversed—another enigmatic point of contention for global warming's true believers.

ICE AGE PANIC

By the mid-1940s, global average temperatures began a big drop of nearly 2°F (1°C) that continued into the seventies. Many scientists at the time were certain we were heading into an Ice Age. In fact, the media was all over it. Here is a statement from the US Congressional Record:

> In 1972, the National Science Board, the governing body of the National Science Foundation, observed: Judging from the record of the past inter-glacial ages, the present time of high temperatures should be drawing to an end... leading into the next glacial age.[5]

---

4   "The Grapes of Wrath," *National Geographic*, April 2014, https://education. nationalgeographic.org/resource/grapes-wrath/.

5   Congressional Record Volume 150, Number 130, Senate, S11291–S11297, Monday, October 11, 2004, https://www.govinfo.gov/content/pkg/CREC-2004-10-11/html/CREC-2004-10-11-pt1-PgS11291-4.htm.

And there are many more similar opinions reflected in head-lines and quotes touting a serious cooldown. Here are a few:

> As for the present cooling trend, a number of leading climatologists have concluded that it is very bad news indeed. *Fortune* magazine, February 1974.

> Climatological Cassandras are becoming increasingly apprehensive, for the weather aberrations they are studying may be the harbinger of another ice age. *TIME*, June 24, 1974.

> There are ominous signs that the earth's weather patterns have begun to change dramatically and that these changes may portend a drastic decline in food production—with serious political implications for just about every nation on Earth. The drop in food output could begin quite soon, perhaps only ten years from now.... *Newsweek*, April 28, 1975.

> Scientists Ponder Why World's Climate Is Changing; A Major Cooling Widely Considered to Be Inevitable. *New York Times*, May 21, 1975.

> The threat of a new ice age must now stand alongside nuclear war as a likely source of wholesale death and misery for mankind. *International Wildlife*, July 1975.

> "The cooling since 1940 has been large enough and consistent enough that it will not soon be

reversed." C. C. Wallen, World Meteorological Organization, *Science News*, 1975.

That the earth's climate changes, and even now might be changing quite rapidly, is widely recognized. The questions facing worried experts are: Is the world as a whole cooling off and perhaps leading to an onset of huge ice sheets? *National Geographic Magazine*, 1976.[6]

And, there is this excerpt from a 1978 investigation conducted by the State of Illinois' Board of Natural Resources and Conservation, entitled *Record Winter Storms in Illinois*:

The Midwest, including Illinois, experienced in 1977–1978 its most severe winter since weather records began in the early Nineteenth Century. Illinois had a record-breaking number of 18 severe winter storms; 4 such storms is normal... The storms led to 62 deaths and more than 2,000 injuries. Utilities, communication systems, and transportation suffered great losses....[7]

During the winter of 1977–78, snowfall records were broken throughout the Midwest and northeastern states, one-hundred-

---

6   Author's note: This is one of the first major articles to ever emphasize the term "climate change," which incites a broader range of fear. To be fair, the next sentence following the quote I employed covers all potential scenarios going forward, stating, "Or are we instead warming our planet irreversibly with our industry, automobiles, and land-clearing practices?" Samuel W. Matthews, "What's Happening to Our Climate?" *National Geographic Magazine*, November 1976.

7   Stanley A. Changnon Jr., David Changnon, *Record Winter Storms in Illinois, 1977–1978*, Illinois State Water Survey, 1978, https://www.isws.illinois.edu/pubdoc/ri/iswsri-88.pdf.

mile-per-hour winds ripped through Indiana, and record cold was experienced across much of the nation. Indeed, similar extreme weather was reported throughout the Northern Hemisphere. In Great Britain, that same winter is memorialized with a line from Shakespeare's *Richard III*, the "Winter of Discontent."[8] The German press was less artist and more absolute in its summary, simply referring to the storms that hit their country as "The Blizzard of the Century."[9]

The next winter, 1978–79, was also intense with the wettest January ever recorded in numerous cities, including Boston, Philadelphia, New York City, and Hilo, Hawaii. Snow records were established or nearly established in parts of Arizona, Wyoming, Kansas, Illinois, Pennsylvania, and New England.[10]

So, it all comes as no surprise that about this same time a bestselling book was published entitled *The Cooling*, by Lowell Ponte, who wrote, "The cooling has already killed hundreds of thousands of people in poor nations."

Interestingly, *The Cooling* was endorsed by a young up-and-coming atmospheric scientist named Stephen Schneider, who is quoted on the book's cover:

> "…this well-written book points out in clear language that the climatic threat could be as awe-

8   James Brigden, "Winters of Discontent: 6 of the Worst Winters in British History," *Sky History*, 2021, https://www.history.co.uk/articles/winters-of-discontent-6-of-the-worst-winters-in-british-history.

9   "Brutal Winter of 1978: 35 Amazing Photos of the Blizzard in Northern Germany from December 1978 to January 1979," *Vintage Everyday*, January 15, 2021, https://www.vintag.es/2021/01/1978-germany-blizzard.html.

10 A. James Wagner, "Weather and Circulation of January 1979," *Bulletin of the American Meteorological Society*, April, 1979.

some as any we might face, and that massive worldwide actions to hedge against that threat deserve immediate consideration...."

With great passion, Schneider advocated "massive worldwide actions to hedge against" global *cooling* (as you will discover in the next chapter, Dr. Schneider would soon flip and become quite the global warmer).

But if you go back another ninety years, one can see that, even then, extreme weather incited anxiety, which is a potent political device. Like a sensational news tease, the more dramatic, the more effective.

In 1883, an eminent political philosopher predicted this calamitous global cool down:

> But inexorably the time will come when the declining warmth of the sun will no longer suffice to melt the ice thrusting itself forward from the poles; when the human race, crowding more and more about the equator, will finally no longer find even there enough heat for life; when gradually even the last trace of organic life will vanish; and the earth, an extinct frozen globe like the moon, will circle in deepest darkness and in an ever narrower orbit about the equally extinct sun, and at last fall into it.[11]

That caustic swipe at the fate of mankind was penned by Marx's writing partner, Frederick Engels. Prominent geologists

---

11 Frederick Engels, *Dialectics of Nature*, "Introduction," (1883), http://www. marxists.org/archive/marx/works/1883/don/ch01.htm.

at that time in 1883 were advancing the theory that the earth was heading into another Ice Age. In fact, twelve years later, on February 24, 1895, a *New York Times* headline declared, "Geologists Think the World May Be Frozen Up Again."

Cooling and warming trends are regular occurrences, but right now the party line is to hide from truth recorded in the past that doesn't align with the present agenda.

HISTORY DOES MATTER

Into the 1990s, it was generally accepted by all corners of the scientific community that over the past one thousand years, Earth's temperature record illustrated a distinct four-hundred-year warming period between AD 900 and 1300. During this span, the average temperature on the planet's surface was likely *at least* 2°F (1.2°C) warmer than today. This is known as the Medieval Warm Period (MWP). In fact, this noteworthy warm stretch was originally recognized—complete with hand drawn graph—in the First Assessment Report of the United Nations Intergovernmental Panel on Climate Change in 1990.

The MWP was followed by a brief temperature stabilization and then a distinct global cooling from AD 1350 to 1800, completely reversing the warmth of the MWP. This 450-year interval is known at the Little Ice Age (LIA), with temperatures at one point falling to roughly 2°F colder than today.

The honest question is, what caused these noteworthy, long-lasting temperature swings? Like the MWP, changes could not be blamed on human industrialization, so what was the driver?

Rather than answer those questions, climate change proponents radically altered the record, as I detail in *Climategate*. The condensed version is that in 1990 the UN's International Panel on Climate Change presented the world with a very simple graph illustrating variations in global temperature since 900 AD. The model made it clear that from 1050 AD to 1350 the average global temperature was significantly warmer than today. Enter a PhD from the University of Oklahoma's geoscience department, David Deming. In 1995, Deming had a research paper published in the journal, *Science*, wherein he presented research based on an extensive study of ancient tree rings. His conclusion stated there seemed to be no cause-and-effect relationship between anthropogenic activities and climatic warming. Despite his conclusion, he remained credible amongst his peers and was included in many group emails with prestigious scientists.

What happened next was the beginning of what many refer to as "the climategate scandal." In a 2008 interview, Deming shared his story with me, explaining that, shortly after his work appeared in *Science*, "A major person working in the area of climate change and global warming sent me an astonishing email that said, 'We have to get rid of the Medieval Warm Period.'" At first Deming didn't think the individual was serious but, as it turned out, there was an intentional effort being made to revise the records. This was confirmed in a large batch of emails that were leaked from the Climate Research Unit (CRU) at the University of East Anglia (UK). Many of the emails revealed what appeared to be a concerted effort to manipulate data and expunge past warming from the records. Eventually, another player, Michael Mann, entered the stage and, through a complex number-crunching exercise, presented a paper in both *Nature* and *Geophysical Research Letters*, which flattened the MWP to nearly nothing. Mann then presented a new graph exhibiting temperatures slightly ebbing and flowing until the twentieth century where they suddenly spike upward like a rocket. It was as if an ice pick had been taken to the frontal cortex of the climate record. The mission was accomplished, and the Medieval Warm Period was dissolved.

Meantime, over the years, Mann has become a vocal progressive activist who pulls no punches regarding his disposition toward those on the opposite side of the climate change debate. In a January 2, 2024, Tweet, Mann stated, "…there is empirical, peer-reviewed support for the conclusion that climate deniers, in general, are truly awful human beings." I responded to Mann's absurd post on my own X account (@debatemealgore), explaining his comments illustrate how those losing an argument "generally become angry." Within an hour after posting, I was blocked from even looking at his feed.

This updated graph flew in the face of copious historical footnotes attesting to the reality and benefits of the MWP. For example, Great Britain is well-known for hearty beer and robust whiskey but certainly not wine. However, during the MWP, England was home to some of the world's finest commercial vineyards. For several hundred years, wines produced by the grapes cultivated in England during that time competed well with those in France and Germany (two regions that continue to produce quality wine to this day). It is important to note, the former vineyards in England are latitudinally several hundred miles *north* of those currently found in France and Germany and are completely incapable of yielding wine grapes today because the climate is simply too cold. Also, back then, German vintners were

CLIMATE CULT

growing grapes at significantly higher elevations than they can today. Remnants of ancient vineyards are visible at elevations of 2,500 feet, whereas Germany's favored wine grapes are currently grown below 1,500 feet. Germany's climate was likely at least 2°F *warmer* than today.

Further botanical evidence from Europe also indicates a previously warmer climate. By observing the Alps, one can clearly see the remnants of ancient trees that once thrived at as much as 1,500 feet above the current tree line. Similarly, present tree lines in Iceland also indicate a frozen forest beneath the sluggish rivers of ice and snow. And there is ample proof that the Medieval Warm Period was a global event:[12] In the Atlantic Ocean's famed Bermuda Triangle, radiocarbon dating of marine organisms in seabed sediments reveals ocean surface temperatures were around 2°F warmer during the MWP than they are today. At Kenya's Lake Naivasha, extracted sediments reveal the lake endured a lengthy two-hundred-year drought from about 1000–1200 AD.[13] In Peru, ice core samples from the Quelccaya Glacier in the Andes reveal there was a distinct "warm, arid period" from 1000–1550 AD.[14] While studying oxygen isotopes in a peat bog in northeastern China near the border of North Korea, Taiwan researchers uncovered a six-thousand-year temperature history,

12 L. D. Keigwin, "The Little Ice Age and Medieval Warm Period in the Sargasso Sea," *Science*, vol. 274 (1996): 1504–1508.
13 D. Verschuren, "Rainfall and Drought in Equatorial East Africa during the past 1,100 Years," *Nature*, vol. 403, no. 6768 (January 27, 2000): 410–414.
14 Lonnie G. Thompson, "Past, Present and Future of Glacier Archives from the World's Highest Mountains," Byrd Polar and Climate Research Center, The Ohio State University, https://www.geo.umass.edu/climate/papers2/Thompson_AmPhilSoc.pdf.

illustrating that from 1100 to 1200 AD, it was about 2°F warmer than it is today.

During that same period, the Vikings were using Greenland as their primary base for operations as they plundered and pillaged much of the world. The snow and ice that once covered the region for much of the year was replaced with grasses and wildflowers. Forests were flourishing, and deciduous trees were taking root and expanding. Excavation of former Viking villages and burial grounds in Greenland have been discovered beneath what is presently permanently frozen soil. Research suggests temperatures were as much as 6.6°F (4°C) warmer back in Viking days.[15] Soil samples show the presence of corn pollen, while growing corn is impossible in Greenland today.

LITTLE ICE AGE

The Medieval Warm Period gave way to a long cooling period from 1350 to 1800. Given that the thermometer was not mass-produced and in widespread use until the 1800s, the conclusions we draw regarding this 450-year period are based on physical evidence and historical observations and clearly illustrate the existence of the period known as the Little Ice Age (LIA). The evidence is overwhelming and includes hundreds of studies demonstrating the physical isotopes of carbon, hydrogen, and oxygen in decayed plants excavated throughout Europe, indicating significantly lower temperatures during the LIA.[16] The examination of tree rings using boreholes indicates a colder climate

---

15   F. Donald Logan, *The Vikings in History*, Second Edition (London and New York: Hutchison and Company, Ltd., 1991).

16   Richard D. Tkachuck, "The Little Ice Age," *Origins* 10-2 (1983): 51–65.

worldwide.[17] Studies of coral, which exhibit their own annual growth bands, yield data that suggests average mean ocean water temperatures cooled significantly during the LIA as well.

Aggressive studies on past climate instituted by independent Chinese researchers unanimously conclude, "China has advantages in reconstructing historical climate change for its abundant documented historical records and other natural evidence obtained from tree rings, lake sediments, ice cores, and stalagmites."[18] The Chinese research points to the LIA being as much as 2°F (1.1°C) cooler than today,[19] matching other global calculations from the same period indicating the Little Ice Age was an event experienced worldwide.

And there were the famous "Frost Fairs," which occurred annually on England's River Thames. Each winter, the Thames froze solidly, often with ice a foot thick (in contrast, the river only froze four times in the twentieth century and *never* a foot thick). The fairs convened in the middle of the river, complete with merry-go-rounds, swings, slippery football matches, and chaotic donkey and horse races. Diarist John Evelyn described activities of the Frost Fairs as "...sleds, sliding with skates, bullbaiting, horse-and-coach races, puppet plays and interludes, cooks, tippling [taverns], and other lewd places, so that it seemed to be a bacchanalian triumph, or carnival on the water."

During the winter of 1620–21, on the other side of the Atlantic, the Pilgrim settlers, who had come to America on the

---

17 Ibid.
18 Ge Quansheng et al., "Key Points on Temperature Change of the Past 2000 Years in China," *Progress in Natural Science*, vol. 14, no. 8 (August 2004).
19 Ibid.

Mayflower in search of religious freedom, had their faith immediately tested by horridly frigid conditions.

In his book, *History of the Plymouth Settlement*, Pilgrim leader William Bradford wrote that within weeks of their arrival in the New Land, "the severity of the winter weather, and sickness, had begun...." Over the next three months, half of their company had died, "partly owing to the severity of the winter, especially during January and February," Bradford reported. "Of all the hundred-odd persons, scarcely fifty remained, and sometimes two or three persons died in a day."

UNTIL NOW

It is imperative to understand that all the material presented in this chapter is critical to the global warming/climate change debate, because activists stubbornly maintain that the Little Ice Age was simply a local event that impacted isolated sectors of the Northern Hemisphere, and they have massaged the Medieval Warm Period into little more than a few anecdotal accounts, providing ample room for Al Gore to speak of the currently "boiling" oceans, droughts "sucking the moisture out of the land," and "waves of climate refugees."

But their blatant scheme is falling apart.

A 2023 paper presented in the *Journal of Geophysical Research: Atmospheres* by a group of NOAA scientists reveals a new satellite-derived temperature record for the global troposphere (the atmospheric layer just above the earth's surface extending upward about eleven miles). The troposphere's climate record is important because it is not affected by urbanization or other changes to the land surface, thus, providing a clean signal of how greenhouse gases could be impacting temperature.

The group started its work more than a decade ago, initially producing a temperature record known as Satellite Applications and Research (STAR). At first, STAR depicted more warming than even RSS. However, based on a new empirical method for removing time-of-day observation drift complications from the satellites and a more stable method of merging satellite records, STAR now reveals a warming in the lower troposphere of 0.09°C (slightly less than even UAH has presented at 0.1°C). On the Fahrenheit scale, this means the current verifiable warming is roughly 0.18°F per decade—*far from alarming*.

For the troposphere, STAR estimates a current warming trend of 0.14°C per decade. This is a severe blow to those predicting climate calamity. These measurements are half the average warming rate predicted by the climate models.

In fact, the NOAA scientists noted their findings were "consistent with conclusions from McKitrick and Christy (2020)."[20] In those results, Ross McKitrick and John Christy state that the climate models routinely used by alarmists "reveal a systematic warm bias."[21]

This is a showstopper. Those in the pews of the climate cult need to hear this powerful, scientific admission from research-

---

20 Cheng-Zhi Zou, Hui Xu, Xianjun Hao, Qian Liu, "Mid-Tropospheric Layer Temperature Record Derived from Satellite Microwave Sounder Observations with Backward Merging Approach," *Journal of Geophysical Research: Atmospheres*, March 3, 2023, https://agupubs.onlinelibrary.wiley.com/doi/full/10.1029/2022JD037472.

21 R. McKitrick, J. Christy, "Pervasive Warming Bias in CMIP6 Tropospheric Layers," *Advanced Earth and Space Science*, July 15, 2020, https://agupubs.onlinelibrary.wiley.com/doi/10.1029/2020EA001281.

ers working within the hallowed halls of the National Oceanic and Atmospheric Administration (during the Biden administration, no less).

The NOAA scientists also say their findings "have strong implications for trends in climate model simulations and other observations."

To add further emphasis, these studies compliment a paper published in 2021 by independent researchers from England and Germany who concluded, "increasing levels of CO_2 will not lead to significant changes in earth temperature."

And there is more.

I happened across a forthright article on NASA's Jet Propulsion Laboratory's website enlightening us to the complexity involved in taking the earth's temperature:

> "Ensuring the accuracy of Earth's long-term global and regional surface temperature records is a challenging, constantly evolving undertaking."[22]

As shown in this chapter, lack of available data and the presence of urban heat islands coupled with human error, bias, and guesstimates can lead to measurement inconsistencies wildly affecting the record. The article from NASA also reveals *estimates* of the actual temperature are often required:

> Scientists have been building estimates of Earth's average global temperature for more than a century, using temperature records from weather sta-

---

22 Alan Buis, NASA's Jet Propulsion Laboratory, "The Raw Truth on Global Temperature Records," *Ask NASA Climate*, March 25, 2021, https://climate.nasa.gov/ask-nasa-climate/3071/the-raw-truth-on-global-temperature-records/.

tions. But before 1880, there just wasn't enough data to make accurate calculations, resulting in uncertainties in these older records.

The NASA post concedes human error is a factor within the earth's temperature recordings: "As all of us know, humans can make occasional mistakes in recording and transcribing observations."[23]

CONCLUSION

The summary of the global temperature record, sans the controversial 1998 magic wand, illustrates:

- At the peak of the Medieval Warm Period (900–1,300 AD), it is estimated the temperature was more than 2°F warmer than today (1.2°C).

- During the Little Ice Age (1350–1800 AD), the warmth of the MWP was wiped out with temperatures falling to 2°F cooler than today (1.2°C).

- Temperatures stabilized and then rose notably, particularly in the 1930s, making it 1°F warmer than today (0.6°C).

- From 1940–1975 there was a slight cooling of 0.18°F (0.1°C).

- A minor warming, 0.34°F (0.19°C), occurred between 1970 and 1998.

23 Ibid.

- According to satellite data, from 1998 to 2014, temperatures were flat.[24]

- Since 2014, temperatures have risen 0.36°F (0.2°C).

No matter the source, average temperatures have risen since the end of the Little Ice Age and the start of the Industrial Revolution, all within historical norms. Climate activists use their charts to say we've dangerously warmed 1.2°C and insist the earth can only withstand an increase of 1.5°C before all hell breaks loose. Many experts rightly disagree with that prediction.

In 2017, a paper written by three well-known dissenters was reviewed by seventeen noteworthy professors and environmental experts whose leanings about anthropogenic climate change are unknown to me. The paper was able to pass muster and was published.[25] The focus of their research investigated the various methods of temperature collection (the data sets previously noted) that are used to derive what is known as the global average surface temperature data (GAST).

Here are their conclusions:

> Clearly the historical GAST data adjustments that have been made have been dramatic and invari-

---

24 "Latest Global Average Tropospheric Temperatures," based on a running, centered, thirteen-month average, Dr. Roy Spencer, PhD, May 2023, https://www.drroyspencer.com/latest-global-temperatures/.

25 Reviewers include Dr. Richard A. Keen, instructor emeritus of atmospheric and oceanic sciences, University of Colorado; Dr. George T. Wolff, former chair, Environmental Protection Agency's Clean Air Scientific Advisory Committee; Dr. Anthony R. Lupo, UN International Panel on Climate Change expert reviewer and professor, atmospheric science, University of Missouri; and Dr. Alan Carlin, retired senior analyst and manager, US Environmental Protection Agency.

ably have been favorable to Climate Alarmists'
views regarding Global Warming.

The conclusive findings of this research are that the
three GAST data sets are not a valid representation
of reality. In fact, the magnitude of their historical data adjustments, that removed their cyclical
temperature patterns, are totally inconsistent with
published and credible U.S. and other temperature data. Thus, it is impossible to conclude from
the three published GAST data sets that recent
years have been the warmest ever—despite current claims of record setting warming. The analysis
above raises grave doubts that any of the GAST
data sets are a credible representation of reality. [26]

Their strong opinions were completely dismissed by the
media, obviously because such investigation does not agree with
the climate agenda. But now, that report, along with the other
research presented in this chapter, should give thoughtful skeptics the confidence to know that "the science" is hardly settled.

From my perspective, the most prolific voice to swim
upstream from the popular narrative is Dr. John Clauser, the
2022 Nobel Prize recipient for physics. Recently, Clauser joined
another Nobel laureate and over 1,600 professionals in signing
the World Climate Declaration (WCD) organized by Climate
Intelligence (CLINTEL). This declaration asserts that there is no

---

26 Dr. James P. Wallace III, Dr. Joseph S. D'Aleo, Dr. Craig D. Idso, *On the Validity
   of NOAA, NASA and Hadley CRU Global Average Surface Temperature Data & The
   Validity of EPA's CO$_2$ Endangerment Finding*, June 2017, https://thsresearch.files.
   wordpress.com/2017/05/ef-gast-data-research-report-062717.pdf.

"climate emergency," that climate change science is not conclusive, and that the earth's history over thousands of years shows a consistently changing climate.

In an article posted at CO_2 Coalition, Clauser states, "I read all of the various [United Nations International Panel on Climate Change] reports [and the] National Academy reports on this. As a physicist, I'd worked at some excellent institutions—Caltech, Columbia, Cal Berkeley—where very careful science needed to be done. And reading these reports, I was appalled at how sloppy the work was. And in particular, it was very obvious, even in the earliest reports, and all carried on through to the present, that clouds were not at all understood.... It's just simply bad science."

Focusing on the weaknesses of the climate models used to forecast climate change, Clauser continued, "I believe I have the missing piece of the puzzle that has been left out in virtually all of these computer programs. And that is the effect of clouds."

He contends such modeling presents something akin to fiction.

"That's a totally artificial Earth. It is a totally artificial case for using a model, and this is pretty much what the IPCC and others use—a cloud-free earth."[27]

This is a huge win for climate truth, which the propagandists suppress from their congregants. Such blatant censorship is emblematic of a cult, intended to perpetuate a continual state of fear, which is needed to keep the masses moving in the proper direction and allow the climate change high priests to continue their indoctrination of humanity.

27 "Nobel Winner Refutes Climate Change Narrative, Points Out Ignored Factor," CO_2 Coalition, September 10, 2023, https://co2coalition.org/news/nobel-winner-refutes-climate-change-narrative-points-out-ignored-factor/.

CHAPTER FIVE

//////////////////////////

FEAR

THE FOLLOWING TRANSCRIPT IS FROM a CNN report on an appearance earlier that day by Al Gore on *Meet the Press*, Sunday, July 24, 2022:

> Former Vice President Al Gore on Sunday likened climate crisis deniers to the police officers in Uvalde who failed to take action as 21 schoolchildren and teachers were gunned down in their classrooms, saying, "They heard the screams, they heard the gunshots, and nobody stepped forward."
>
> "The climate deniers are really in some ways similar to all of those almost 400 law enforcement officers in Uvalde, Texas, who were waiting outside an unlocked door while the children were being massacred," Gore told NBC's *Meet the Press*. "They heard the screams, they heard the gunshots, and nobody stepped forward."[1]

---

1   Sarah Fortinsky, "Al Gore Compares Climate Deniers to Uvalde Police Who 'Heard the Gunshots, and Nobody Stepped Forward,'" CNN Politics, July 24, 2022, https://www.cnn.com/2022/07/24/politics/al-gore-climate-deniers-uvalde-police/index.html.

What the world heard in that news report was an untethered, vicious labeling of anthropogenic climate change skeptics. In verbal order, the CNN news writers and the former vice president of the United States implied that such individuals are deniers, cowards, and guilty of something akin to aiding and abetting a mass murder. This kind of messaging is part of a blatant scare campaign representative of a comprehensive, compulsory indoctrination program being employed via all popular communication mediums and in every school in the developed world. It does not matter how many predictions of doom fail to manifest, the false prophets spouting these lies are never held accountable.

This rhetoric is often referred to as brainwashing, greenwashing, or even mind control; psychologists refer to these methods of coercion as "fear appeals." Unprecedented pressure is being placed upon the masses to dogmatically believe that humanity is on the brink because of lifestyle choices dependent on energy derived from fossil fuels. Carbon dioxide, once taught to be life's essential ally, is now the enemy. Our planet's history of fluctuating temperatures has been blotted out and replaced with sensationalized graphs that spike upward in the present like a red-hot skewer. Alternative deliberation is not tolerated.

Through blatant media bias, academic indoctrination, psychological intimidation, celebrity endorsements, and an increasingly extortionate business culture, we have come to this time and place whereby so many have been easily drawn into the climate cult.

YELLOW JOURNALISM

In 1864, a young man immigrated to the United States from Hungary, settling in St. Louis, Missouri. He took up work at a tavern, where he became influenced by Henry Clay Brockmeyer, a flamboyant local philosopher, poet, and state senator who was leader of the local Hegelian club. You will recall that Karl Marx was also a Hegelian club member; Marx's branch had a goal to "liquidate Christianity." It is said that the young immigrant "would hang on to Brockmeyer's thunderous words even as he served them pretzels and beer."[2] Our subject was quite taken by Brockmeyer's wide-ranging discourses that included Hegel's unorthodox social, economic, and political sentiments. The young man eventually became a newspaper reporter as well as a mover and shaker in Missouri politics, eventually being elected to the state's legislature. He became good friends with political cartoonist Joseph Keppler, publisher of *Puck* magazine and known in history for his powerful, one-panel "Seeds of Socialism" cartoon in which a giant boar is seen holding the dome of the US Capitol while trampling an American flag and sowing seeds onto a field sprouting "socialist votes." In 1878, our subject purchased the *St. Louis Post Dispatch*, and in 1883, he moved to New York to buy the lackluster *New York World* newspaper with hopes of making it the voice of the Democratic Party, announcing that the paper would be "dedicated to the cause of the people rather than that of purse-potentates."[3]

New York World quickly turned around as our man trained his staff in a writing and reporting technique that teased read-

---

2   W. A. Swanberg, *Pulitzer* (Scribner, 1967), 6.
3   W. A. Swanberg, *Pulitzer* (Scribner, 1967), 70.

ers with flashy headlines, sensational stories, scandals, horrors, dramatic illustrations, eye-popping graphics, and reporting that was often unsubstantiated, light on the facts, and dripping with political and social commentary. By 1900, his paper was tops in the country. *New York World's* brand of journalism took reporting from rational and informative to emotional and sensational. The distinct style would be known as "yellow journalism."

The publisher was soon the toast of the town and in 1884 was elected to the US House of Representatives. However, realizing he had more political influence running his newspaper, he resigned from Congress before his first term was completed.

What is ironic about this man is that in 1892, he approached the president of New York City's Columbia University with an offer to fund the world's first school of journalism and establish an annual prize for the press—perhaps it was to improve his image as a legitimate keeper of the fourth estate, or maybe it was his hope that such a school could produce the next generation of radically biased reporters. Columbia turned him down. However, following his death in 1911, this newspaper publisher, who had become one of the wealthiest men in America with an estate worth $30 million, bequeathed $2 million to Columbia University. A journalism school was established, and the new prize for journalism was announced: the Pulitzer Prize, named in honor of the man who wrote the book on media sensationalism, Joseph Pulitzer.

Over the following decades, yellow journalism strategies gradually became mainstream as the press transformed itself into a stage on which the actions of government and society are presented in a series of gripping dramas. Pulitzer's initial news template gave us the current theatre of information we are inundated with today. Special interest groups, activists, lobbyists, and those with subversive agendas now provide the many performing stages using countless actors and endless conflicts. In the case of global warming and climate change, it is my contention that Pulitzer's influence has steadily provided that agenda with a secure place of residence, going back to headlines associated with the first Earth Day in 1970:

> "It is already too late to avoid mass starvation."
> —Earth Day co-founder Denis Hayes,
> *The Living Wilderness*, spring 1970.

> "Dr. S. Dillon Ripley, secretary of the Smithsonian Institute, believes that in 25 years, somewhere between 75 and 80 percent of all the species of living animals will be extinct."
> —Earth Day co-founder Senator Gaylord
> Nelson, *Look* magazine, April 1970.

> "It took several million years for the population to reach a total of two billion people in 1930, while a second two billion will have been added by 1975! By that time, some experts feel that food shortages will have escalated the present level of world hunger and starvation into famines of unbelievable proportions. Other experts, more optimistic, think the ultimate food-population collision will not occur until the decade of the 1980s…Make no mistake about it, the imbalance will be redressed."
> —Earth Day co-founder Paul Ehrlich,
> writing the foreword to *Nightmare Age*
> by Frederik Pohl, Ballantine Books,
> 1970.

"Man must stop pollution and conserve his resources, not merely to enhance existence but to save the race from intolerable deterioration and possible extinction."
—*New York Times* editorial, April 23, 1970.

Stephen Schneider, who provided the cover endorsement for *The Cooling*, would go on to become a celebrated professor of environmental biology and global change at Stanford University. Thirteen years after *The Cooling*, he flipped his cards like a magician and released his own book, *Global Warming*. In association with the book's release, Schneider did an interview with *Discover* magazine, openly sanctioning reckless environmental hyperbole:

[W]e are not just scientists but human beings as well. And like most people, we'd like to see the world a better place, which in this context translates into our working to reduce the risk of potentially disastrous climatic change. To do that we need to get some broad-based support, to capture the public's imagination. That entails loads of media coverage. So we have to offer up scary scenarios, make simplified dramatic statements, and make little mention of any doubts we might have.[4]

"Broad-based support...loads of media coverage...scary scenarios...dramatic statements." Schneider presents us with the perfect blend of science and yellow journalism. It should not

4   Stephen Schneider, *Discover*, October 1989, 44–45.

CLIMATE CULT

surprise anyone that Dr. Schneider would go on to be known as a climate activist who authored another book, *Science as a Contact Sport*.

EDUCATION OR INDOCTRINATION?

In addition to the media's role in intentionally providing climate activists with loads of misleading coverage, our educational system essentially works as a synchronized incubator for the agenda. I know we all want to believe our local schools are doing an outstanding job, and I recognize there are some phenomenal teachers in these schools. However, when it comes to climate change, instruction has been replaced with shameless indoctrination.

There are over thirteen thousand independent public-school districts in the US, each generally overseen by locally elected school boards. Curriculum is most often chosen by professional staff in the district or the state and then presented to the school board for approval. This is why many contend the most influential elected officials in the country are members of the school board. It is no wonder why so many progressives pack these boards.

Now, enter the Progressive Change Campaign Committee (PCCC), which bills itself as "fighting on democracy issues." They have launched an aggressive nationwide campaign called "Save Our School Boards."[5]

Simultaneously, the PCCC is fighting for the climate agenda. The organization contends, "Climate change is an existential threat...Economic inequality, racial injustice, and the cli-

5   Progressive Change Campaign Committee, About Page, https://www.boldprogressives.org/about/.

mate crisis are not different problems, they are part of the same problem."[6]

To the PCCC, not only is climate change an "existential threat," but "economic inequality and racial injustice" are one and the same. This is the message they want to make sure every school board in America promotes: climate change is an immutable fact affecting the social issues of our day. That decree resounds with progressives in the arena of education. For example, the Next Generation Science Standards (NGSS), founded in 2011 under the direction of Achieve, Inc. (the same organization that developed the controversial Common Core national standards for English, arts, and mathematics), are used in countless schools across America. I examined the NGSS expectation guidelines for the environmental and climate curriculum. After studying climate change, middle school students are expected to agree with these verbatim statements:

- The major role that human activities play in causing the rise in global temperatures.[7]

- Human activities have significantly altered the biosphere, sometimes damaging or destroying natural habitats and causing the extinction of other species.[8]

- Human activities, such as the release of greenhouse gases from burning fossil fuels, are major factors in the cur-

---

6   Progressive Change Campaign Committee, Our Issues, "Green New Deal," https://www.boldprogressives.org/issues/green-new-deal/.

7   Next Generation Science Standards, MS-ESS3-5 Earth and Human Activity, Achieve, Inc., 2017, 71.

8   Next Generation Science Standards, ESS3.C Human Impacts on Earth Systems, Achieve Inc., 2017, 71.

rent rise in Earth's mean surface temperature (global warming).[9]

In accordance with the NGSS standards, high school students should come away from the curriculum agreeing:

- Human activity is also having adverse impacts on biodiversity through overpopulation, overexploitation, habitat destruction, pollution, introduction of invasive species, and climate change.[10]

- Changes in the atmosphere due to human activity have increased carbon dioxide concentrations and thus affect climate.[11]

- Current models predict that, although future regional climate changes will be complex and varied, average global temperatures will continue to rise.[12]

- An example of the far-reaching impacts from a human activity is how an increase in atmospheric carbon dioxide results in an increase in photosynthetic biomass on land and an increase in ocean acidification, with resulting impacts on sea organism health and marine populations.[13]

9   Next Generation Science Standards, ESS3.D Human Impacts on Earth Systems, Achieve Inc., 2017, 71.

10 Next Generation Science Standards, LS4.D Biodiversity and Humans, Achieve Inc., 2017, 90.

11 Next Generation Science Standards, ESS2.D Earth's Systems, Achieve Inc., 2017, 99.

12 Next Generation Science Standards, ESS2.D Earth and Human Activity, Achieve Inc., 2017, 100.

13 Next Generation Science Standards, HS-ESS3-6 Earth and Human Activity, Achieve Inc., 2017, 100.

Nowhere in the NGSS curriculum are teachers equipped to test or challenge the hypothesis of anthropogenic global warming or climate change. Instead, students are repeatedly told throughout their education that climate change is an inarguable fact, caused by greedy human behavior. Instead of introducing students to the world of scientific inquiry, NGSS seeks to inculcate progressive social values by engaging students during classroom instruction, persuasively pressuring them to become active participants in rescuing the planet in accordance with the environmentalist gospel.

This is not education; it is indoctrination loaded with heavy doses of anxiety. This is the same thing the media is guilty of. These are conspired *fear appeals*.

FEAR APPEALS

According to the National Library of Medicine:

> Fear appeals are persuasive messages that attempt to arouse fear by emphasizing the potential danger and harm that will befall individuals if they do not adopt the messages' recommendations.[14]

An article in *Psychology Today* expands on this definition, with extremely specific warnings for those who employ fear appeals:

> Clearly, fear appeals do not only provoke fear reactions. There can be feelings of disgust, anger, anxiety, or guilt. These emotions may also have

---

14   Melanie B. Tannenbaum et al., "Appealing to Fear: A Meta-Analysis of Fear Appeal Effectiveness and Theories," National Library of Medicine, National Center for Biotechnology Information, 2015.

an effect on behavioral changes... The practice
of fear appeals has its warranted critics; especially,
if applied unethically. If fear is used to motivate
attitudinal and behavioral change, it should be
used judiciously. Moreover, it might be best
directed toward individuals who are known and
for whom the recommendations apply appropriately, rather than an ambiguous population.[15]

So, when a popular socialist politician in the United States,
who has attained rock star status in the eyes of millions of voters, regularly utilizes fear appeals to gain support for her party's
climate agenda, is she doing so within guidelines put forth by
mental health professionals? Check out these quotes:

"Our planet is going to hit disaster if we don't
turn this ship around and so it's basically like,
there's a scientific consensus that the lives of children are going to be very difficult.

And it does lead, I think, young people to have a
legitimate question, you know, 'Is it okay to still
have children?'"[16]

"We talk about existential threats, the last time
we had a really major existential threat to this
country was around World War II, and so we've

15 Shoba Sreenivasan, PhD, and Linda E. Weinberger, PhD, "Fear Appeals, an
Approach Used to Change Our Attitudes and Behavior," *Psychology Today*,
September 18, 2018, https://www.psychologytoday.com/us/blog/emotional-
nourishment/201809/fear-appeals.
16 Rep. Alexandria Ocasio-Cortez, Instagram Live video, February 24, 2019.

been here before and we have a blueprint of doing this before."[17]

"We don't have time to sit on our hands as our planet burns. For young people, climate change is bigger than election or re-election. It's life or death."[18]

The above quotes were made by Democratic congressional representative Alexandria Ocasio-Cortez, popularly known as "AOC." She is, in my opinion, guilty of dangerous violent messaging that could very well harm a person's emotional state.

Empirical research on fear appeals has resulted in a body of evidence illustrating that fear-arousing communications have a strong effect. Intense fear messages can alter individuals' attitudes, intentions, and behavior—the stronger the messaging, the more significant the emotional consequences. Research also reveals the most crucial factor in changing an individual's actual behavior is perceived vulnerability—people will take action when they feel their back is against the wall, especially if it involves their well-being.[19]

Again, AOC's statement: "We don't have time to sit on our hands as our planet burns…It's life or death."

This is the type of negligent speech that could cause someone to overreact. Yet the climate agenda strategically and routinely utilizes similar fear tactics.

17 Ibid., Town Hall, Queens, New York, October 18, 2018.
18 Ibid., Twitter, @AOC, December 20, 2018.
19 Psychology, IResearchNet, "Fear Appeals," January 10, 2016, https://psychology.iresearchnet.com/social-psychology/emotions/fear-appeals/.

For three generations since the '70s, our politicians, schools, and news media have been feeding us a steady diet of green fear appeals, designed to frighten the masses into resolute belief, submission, and compliance. With every heat wave, tornado, hurricane, lightning-induced forest fire, and blizzard, news coverage on extreme weather events gives audiences the impression such occurrences are new to the planet, especially when the script instantly aligns these otherwise natural events with human-caused climate change. It has become so ridiculous that TV newscasts now regularly hype the "feels like" temperatures to juice up their hot weather coverage. The "feels like" is technically known as the heat index, which combines air temperature and humidity in an attempt to determine how warm it *feels* to a person outdoors. When humidity is high, sweat cannot evaporate as easily, so people have a harder time cooling off. For example, a 90°F temperature combined with 65 percent humidity gives a "feels like" reading of 103°F. Though wind can provide a cooling sensation, the index does not include that. Nonetheless, the "feels like" readings bring customers into the climate cult tent, ready to take action. There is a big push right now in the TV weather biz to use the heat index in the summer and its counterpart, the windchill, in the winter as the primary reading presented to the public.

The biggest and most awesome show on earth—weather—is being used to frighten us into bowing to the globalist demands, ultimately to deconstruct the United States. The fear appeals are working so well that even corporate America has succumbed to the pressure to conform.

ESG RATINGS

Following the Supreme Court's *Dobbs* decision rolling back the *Roe v. Wade* abortion laws, major corporations affirmed they would cover employee travel costs to abortion clinics in states where the procedure is still legally allowed, while other companies refused to acquiesce. Protests and boycotts continue to fly in both directions as consumers reveal their consciences with their wallets. But pro-abortion measures aren't the only way corporations are politicizing themselves. Businesses are actively being judged by their ESG scores: environment, social, and governance. It is a cultural Marxist shakedown operation declaring profit is no longer the only measure of success.

"E" is not about being a good steward of the environment; it's about "climate justice." Based on the assumption that the world is fast approaching disaster, climate justice implies the goal of every organization is to be net zero. For-profit and non-profit corporations must do everything possible to reduce their carbon footprint, no matter how costly it may be. And unless that company wants to suffer the consequences of fierce public opposition, they must fall in line. Like sustainable development (which is exposed in the next chapter), climate justice is multifaceted. According to the United Nations Children's Fund (UNICEF), climate justice also means "combating social injustice, gender injustice, economic injustice, intergenerational injustice and environmental injustice."[20]

---

20 Cristina Colón, "What Is Climate Justice? And What Can We Do to Achieve It?" UNICEF, https://www.unicef.org/globalinsight/what-climate-justice-and-what-can-we-do-achieve-it.

"S" provides scoring metrics for gender and diversity inclusion, relationships with labor unions, mental health of employees, and community relations. "S" is a slap in the face to the for-profit corporate world. Generally, those who risk their own capital by starting or funding a business are the ones who have the first opportunity for greater reward. But not with ESG. The management team running a company works not just for the shareholders but also for stakeholders who are given a voice in the boardroom, even though they have no skin in the game. Stakeholders have a powerful influence on decisions, often at the expense of the shareholders who retain personal financial risks and obligations.

"G" is for governance and applies to the board of directors, their race and gender makeup, political contributions, lobbying, hiring diversity, employee benefits, and community philanthropy.

ESG, then, is an imposed measurement of things that do not relate to business performance, and, because living up to such standards is costly, ESG principles tend to hinder new players from entering the market. Startups find themselves pressured for disclosure of everything from their staff's race and gender to every bit of carbon dioxide expelled from their supply chain. Their survival is dependent on deep-pocketed investors.

Of course, ESG is another subversive plot concocted by the United Nations. It was first presented in a 2004 report, "Who Cares Wins: Connecting Financial Markets to a Changing World." The paper was a joint effort by twenty financial institutions highlighting proposals to their industry on how to better apply environmental issues, liberal social concerns, and politically correct corporate governance standards into their business practices. The recommendations were applauded by the World Economic

Forum's Davos devotees, and soon there was a scramble to comply. Now, according to Bloomberg, global assets for ESG-related exchange traded funds surpassed $35 trillion in 2020 and could reach $50 trillion by 2025.[21]

While many consider ESG to be silly, virtue signaling, woke capitalism, or "greenwashing," there is no denying that companies are falling in line, as institutional investment firms such as banks, mutual funds, hedge funds, insurance companies, and pensions seek compliance. A survey touted by US Bank reveals more than two-thirds of investors (67 percent) believe they have a responsibility to invest in companies that "make the world a better place," while more than half (51 percent) say they avoid investing in companies that don't align with their personal values.[22] According to the survey, the companies that investors want to avoid are, in order: tobacco, private prisons, oil, weapons manufacturers, casinos, "companies that donate to a political party/ politicians I disagree with," and cannabis. The same study reveals the investors most concerned about "socially responsible investing" are Generation Z, followed by registered Democrats. No surprise, especially since Gen Z has lived in a fishbowl of indoctrination and propaganda, and the Democrats' preferred trusted source for news is CNN.[23]

---

21 Press Announcement, "ESG May Surpass $41 Trillion Assets in 2022, but Not without Challenges, Finds Bloomberg Intelligence," *Bloomberg*, January 24, 2022, https://www.bloomberg.com/company/press/esg-may-surpass-41-trillion-assets-in-2022-but-not-without-challenges-finds-bloomberg-intelligence/.

22 Julie Ryan Evans, "51% Avoid Investing in Stocks That Don't Align with Personal Values," *Magnify Money*, June 19, 2021, https://www.magnifymoney.com/news/socially-responsible-investing-survey/.

23 *The Economist*/YouGov Poll, March 26–29, 2022: 1500 US Adult Citizens, https://docs.cdn.yougov.com/3ixnq9227y/econTabReport.pdf#page=224.

The only way publicly traded businesses in those sectors can improve their ESG ratings is by going over-the-top—making efforts to reduce their overall carbon footprint, enacting a strong pro-abortion policy, sponsoring the pride parade, hiring union labor, making diversity appointments on the board of directors, and giving ear to activist organizations—in order to boost an ESG profile. Entities deemed unworthy of a high ESG score, such as oil and gas companies, firearm manufacturers, or the meat and dairy industry with its high levels of methane flatulence, will have to purchase so-called carbon credits to make up for their deficiencies (carbon credits are purchased from companies who promise to plant a tree or invest in renewable energy to balance out whatever carbon emissions are generated by the buyer's business). While there are no uniform standards for how to accurately calculate any actual offsets created by these credits, they are bought and sold via online exchanges and then used to boost ESG profiles. Or these unfavored sectors could follow the lead of Nike and Apple, companies that claim to be all about diversity and inclusion and are extremely sensitive to their carbon footprint, yet they're able to manufacture their products in countries that ignore pollution and human rights violations.

ESG is being utilized to put teeth into the climate agenda, promote a progressive morality code, hijack capitalism, endanger the retirement savings of the middle class, and promote corporate shake-ups. Enter the next phase of corporate compliance: impact investing.

IMPACT INVESTING

Christopher James is a wealthy San Francisco money manager, controlling multibillion-dollar funds investing in health care, tech, and energy. But when his children asked how he could make money in fossil fuels and still be an environmentalist, he didn't have a satisfactory answer.

"I was making my money in tech and health care, just like I was making my money in energy, but I had too much compartmentalization to feel good about what I was doing," James said in an interview with *Harvard Business School*.[24]

James's way of feeling good about himself was to shake up one of the largest oil companies in the world, ExxonMobil. In December 2020, he launched a hedge fund called Engine No. 1, naming it after a fire station down the street from his office. He made a $40 million investment in Exxon—0.02 percent of the company—to go after the "E" in Exxon's low ESG rating. James's goal was to replace four of Exxon's twelve board members with people who had experience with renewable energy. The strategy is now widely known as "impact investing."

Going after ExxonMobil seemed easier than targeting other big oil players, as the company had endured a year of red ink and had yet to cave to pressure on environmental issues. James launched an aggressive PR campaign and networked with other sympathetic shareholders. At ExxonMobil's annual meeting on May 26, 2021, dissidents in the room began making Engine No. 1's case for change. Things became so heated that halfway through

---

24 Lane Lambert, "ESG Activists Met the Moment at ExxonMobil, but Did They Succeed?," Harvard Business School, February 16, 2023, https://hbswk.hbs.edu/item/esg-activists-met-the-moment-at-exxon-mobil-but-did-they-succeed.

the meeting, Exxon's CEO called a time-out to rally for cooler heads to prevail. In the end, three of James's four candidates won board seats. It was a major victory for the climate agenda.

Boycotting certain energy investment is a subset of ESG-related goals. ESG mandates spearheaded by lawmakers in DC are a part of the reason the United States now produces six hundred thousand fewer barrels of oil per day than at the pre-COVID peak, despite rising demand and oil prices hovering near historical highs. Americans should be enjoying an energy renaissance complete with affordable prices both at the pump and on the home utility bills, but that is hardly the case. Meantime, $53 billion in federal fuel tax revenue will roll into the government coffers this year.

And here is something rarely mentioned—aspects of ESG ratings and impact investment may well be illegal. Since the passage of the 1890 Sherman Anti-Trust Act, "Every contract...or conspiracy, in restraint of trade or commerce among the several States, or with foreign nations, is hereby declared to be illegal."[25] Certain ESG proponents may openly be conspiring to restrain trade and commerce by constricting the flow of investment into particular sectors of the economy, especially regarding oil, gas, and coal. Republicans in the US House of Representatives Judiciary Committee began looking into this in December 2022—and they need to complete the job.

One entity that has been specifically called out regarding these potential violations is Climate Action 100+, which states it exists to "ensure the world's largest corporate greenhouse gas

25   Sherman Anti-Trust Act, Section 1 (1890).

emitters take necessary action on climate change."[26] Among the organization's signatories is the world's largest asset manager, BlackRock. Says a member of Climate Action's steering committee, "Given BlackRock's size and influence, their commitment to accelerating engagements with the largest corporate greenhouse gas emitters on climate change sends a powerful signal to companies to reduce emissions, improve corporate governance, and strengthen their disclosure."[27]

Not only is this an example of impact investing on steroids, but Climate Action admits their actions have had "an enormous impact and generated ground-breaking net-zero emission targets from some of the highest-emitting companies around the world."[28]

When companies agree to work together to punish disfavored industries in an effort to advance their ESG, the coordinated activities may violate the antitrust laws and, thus, harm American consumers. Meanwhile, according to the former attorney general of Arizona, Mark Brnovich, Wall Street firms are on the record for "bragging about their coordinated efforts to choke off investment in energy."[29]

---

26 Ceres, "BlackRock Joins Climate Action 100+ to Ensure Largest Corporate Emitters Act on Climate Crisis," January 9, 2020, https://www.ceres.org/news-center/press-releases/blackrock-joins-climate-action-100-ensure-largest-corporate-emitters-act.

27 Ibid.

28 Ibid.

29 Mark Brnovich, "ESG May Be an Antitrust Violation," *Wall Street Journal*, March 6, 2022, https://www.wsj.com/articles/esg-may-be-an-antitrust-violation-climate-activism-energy-prices-401k-retirement-investment-political-agenda-coordinated-influence-11646594807.

YOUR PERSONAL ESG

The big question is, how long will it be before you are judged by your own personal ESG score?

Sooner than you might think.

In 2021, the investment firm Merrill, a division of Bank of America, added a new feature assigning customers a personal ESG score based on their investment portfolios, even for those whose value is under $1,000. One analyst at the influential credit score firm, FICO, says, "Over the longer term, we expect that ESG and climate risk evaluations will become an integral element of credit risk and affordability assessments, and banks and financial institutions will increasingly seek to help consumers and businesses to improve their carbon footprint through provision of education, insights and incentives."[30]

Those three terms—*education, insights*, and *incentives*—are disturbing.

What they really mean is reeducation, surrendering to phony theories, and heavy-handed arm-twisting. Look no further than the current plans the Democratic Party has for consumers. In the name of climate change, natural gas stoves, furnaces, and water heaters are on the brink of extinction, to be replaced exclusively by electric versions. The fantasy of net zero will also require air conditioning units, refrigerators, washing machines, dishwashers, and even current home backup generators to be replaced. In addition to costing the consumer a lot, the new all-electric versions will cause the future decarbonized grid to be constantly in rolling blackout mode. The Democrats' plans, however, will

---

30 Doug Craddock, "Lending Predictions 2022: From BNPL to ESG (and More)," FICO Blog, https://www.fico.com/blogs/lending-predictions-2022-bnpl-esg-and-more.

be a boon to appliance manufacturers like General Electric and LG, both deemed to be in good standing with the WEF, not to mention Whirlpool, who formalized a commitment to the UN "to uphold sustainable and responsible business practices in day-to-day global operations."[31]

For those who refuse to go along with the Left's program, the consequences will be harsh. One could get frozen out of financial markets and denied access to banks, credit and debit cards, loans, and insurance policies. Finding employment also could be challenging. And, if physical cash is banned and money digitized (as the WEF proposes), ESG rebels will find themselves having to barter for goods and services.

The goals of ESG will redesign the free marketplace for the worse and transform the global economy by limiting the supply of goods that consumers demand while deliberately coaxing them to purchase alternative products.

A chilling sign that *everyone* should consider occurred in 2021, when Deutsche Bank AG and Signature Bank announced that they would no longer provide financial services to former president Donald Trump or his business, the Trump Organization, purely for ideological reasons. We are, without a doubt, entering a new reality where individuals are beginning to be judged by their personal ESG score.

Sarah Norman, the person in charge of ESG thought leadership for Merrill and Bank of America, says, "Millennial and Gen Z investors are especially attuned to the idea of stakeholder

31 "Whirlpool Corporation CEO Marc Bitzer Signs onto United Nations Global Compact," PR Newswire, February 19, 2020, https://www.prnewswire.com/news-releases/whirlpool-corporation-ceo-marc-bitzer-signs-onto-united-nations-global-compact-301007217.html.

capitalism versus shareholder capitalism."[32] Stakeholder capitalism contends that businesses should be assessed on whether their actions also benefit both the environment and the diverse communities they serve. Because younger generations have been prepped on these issues, they are eager to buy in. Bank of America is one of the many financial institutions offering customers the ability to align their portfolio with companies that maintain a high ESG rating.

The Impact Investor website states that your ESG score is already being used to hold you accountable for your actions:

> A Personal ESG Score evaluates an individual's performance and impact based on three main factors: environmental, social, and governance…a personal ESG score offers insight into a person's commitment to sustainable practices and responsible decision-making. The purpose behind personal ESG scores involves promoting more mindful behaviors towards the environment and society. By holding individuals accountable for their actions, these scores encourage better choices and habits, leading to a positive change on a larger scale.[33]

And companies like Merrill say they are ready to help you boost that rating. "By looking across three broad categories—People, Planet, and Principles of Governance—we can help

---

32 "Why the 'S' in ESG Is Still in Focus," Merrill, September 16, 2022, https://www.sit1.ml.bac-assets.com/articles/why-social-is-important-to-esg-investing.html.

33 "How to Calculate Your Individual ESG Score," The Impact Investor, June 22, 2023, https://theimpactinvestor.com/calculate-individual-esg-score/.

you choose investing strategies that drive positive change in the world."[34]

These are investment strategies that will determine winners and losers. Never mind the losers could someday be otherwise credit-worthy individuals trying to obtain a car loan, a mortgage, a credit card, or open a bank account. A low personal ESG might well keep someone from employment opportunities if his low rating is considered harmful to the company's average employee ESG valuation. Anna Snider, head of due diligence in the Chief Investment Office for Bank of America, boasts, "Now, individuals can direct their investment dollars to join the movement to harness trillions of dollars in capital toward creating a more sustainable future for us all."[35]

Creating a more sustainable future for us all!

So nice, neat, and admirable, until you examine the fine print.

CONCLUSION

Climate change believers rarely spout anything more substantial than rationalizations based on groupthink and upheld by 24/7 media bombardment, incessant educational indoctrination, and corporate faux altruism. This lack of personal knowledge often leads to variations of public virtue signaling or attempting to claim the moral high ground on controversial issues. Like the hedge fund manager, it makes the signaler "feel good" they

34 "Sustainable and Impact Investing," Merrill, https://www.ml.com/solutions/impact-investing.html.

35 "The Roadmap for Investing in a Sustainable World," Bank of America, December 12, 2022, https://about.bankofamerica.com/en/making-an-impact/sustainable-finance.

are doing the right thing and have the correct answers on complex issues.

Virtue signaling is also a way to call out an individual, company, or organization for backing an idea the signaler disagrees with. It's really an extension of the cancel culture—a politically correct attempt to stifle someone else in an effort to demonstrate personal ESG. The facts don't really matter to the virtue signaler. It's as easy as an emoji, meme, bumper sticker, or purchasing carbon offsets for your next flight to an air-conditioned resort in Hawaii. This is exactly what I believe the climate agenda desires: play this little game of collective thinking, and gradually—mindlessly—people will submit to their schemes.

We've entered an era where few seem interested in honest, genuine debate on cultural and political issues. Rather than engage in a civil discussion, those in the climate change cult have been trained to respond to slogans, something the cult leaders have mastered.

CHAPTER SIX

BUSINESS PLAN

THE COMPANY'S ESG ADOPTION ASIDE, Nike possesses an excellent name for a sports brand: the Greek goddess of victory. And their slogan, *Just Do It*, evokes immediate excitement and a call to action. Likewise, the driving goal of the retail chain Target is instantly apparent: one place where you can find it all. Target's slogan perfectly complements its name: *Expect More. Pay Less.*

Brilliant.

And in a similar vein, environmentalists have done an effective job when it comes to marketing their plans and slogans. As originally employed, the moderately canny term *global warming* was conceived to describe a corrupted atmosphere supposedly becoming unbearably warmer because of fossil fuels. However, for those living in regions of the world that annually experience several months of wintry snow and ice, global warming did not elicit the sense of doom and gloom the green branders had hoped for. However, when they began to refer to the supposed problem as *climate change*, well, now they were on to something that moved the fear needle. Hot weather, cold weather, storms, sea level, snowpack, glaciers, and even the four seasons themselves were appropriated to make the argument that climate change was artificially spiked by selfish human behavior.

But even global warming/climate change still needed a jazzy slogan or snappy catchphrase—one that could neatly represent the intentions of the overall agenda.

Sustainable development.

While it doesn't roll off the tongue, it is a metaphoric term that can conjure up quite a few things. For example, ask a random group of friends what comes to mind when they think of sustainable development, and there will be many different answers.

"Recycling."

"Solar and wind energy."

"No more fossil fuels."

"Organic farming."

"Electric cars."

"Protecting nature."

Sustainable development evokes personal investment. It sparks rhetorical illusion, which, on the surface, sounds great to the rank and file. It calls to mind a myriad of intentions—as anticipated.

"Yeah, sustainable development, protecting the environment—I'm in!"

However, ask people who have been coached on the real meaning behind the phrase, and the answers will be significantly more refined.

"Equity."

"Social justice."

"The end of capitalism."

"Meeting the needs of the present without compromising the future."

While the first set of responses are fruits of sustainable development (recycling, solar and wind, and so on), the latter are the literal intentions: equity, social justice, the abolishment of capitalism, and an uncompromising plan for humanity's future.

Put another way, climate change is the illness brought about by greed, development, profit, and the American creed of life, liberty, and happiness. Sustainable development is the antidote, promising to right the perceived wrongs, determining fairness, and forcing a philosophical transformation of society. ESG is the practical application of sustainable development; and like ESG, sustainable development was created by the United Nations.

THE UNITED NATIONS: A PUNCHLINE

The United Nations was founded following World War II, in 1945, at a meeting in San Francisco with representatives from fifty-one countries in attendance. According to the UN Charter, its primary goals are to:

- Maintain international peace and security

- Develop friendly relations among the nations

- Achieve international cooperation in solving international problems of an economic, social, cultural, or humanitarian character

Obviously, the United Nations' first goal—peace and security—has failed miserably. The organization has since grown to 193 member states whose representatives and bureaucrats operate in a thirty-nine-story building in Manhattan. The foreign workers live very nicely in nearby posh neighborhoods, relish their diplomatic immunity, and are pleased as punch to be living away from their home countries.

The UN is notorious for its many words and endless paper shuffling, yet, despite their heralded charter, they have never stopped, let alone prevented, a war. Over one-quarter of the countries represented at the UN are essentially dictatorships, several of whom even sit on the UN Human Rights Council.

Their real goal? World socialism, communism, and authoritarianism.

Allow me to get personal. My grandfather escaped Lenin's Soviet Russia as a teenager. His family stayed behind and eventually perished at the hands of tyranny.

Shortly after the Soviet Union finally collapsed in 1991, I had the life-changing experience of meeting with three, thirty-something pastors from Russia who were in the US for a church conference. Between them, they had spent over three decades in horrid prisons because of their Christian faith. Their testimonies were stirring.

I have met many from Cuba whose assets were stolen by the communists during the takeover in the 1950s. These wonderful people fled for America and now cherish life, liberty, and personal property.

I was recently introduced to a woman who spent ten years in a Chinese concentration camp simply because of her anti-communist opinions. She eventually escaped and lived underground in her own country until she was finally able to sneak out and eventually find freedom in America.

A gentleman who helped proofread the original manuscript for this book is a former combat Navy Seal who has personally faced the military apparatus of communist regimes. He understands their underhanded determination to see America fail.

I met "Sami" in Eastern Europe. He is Persian, originally from Iran, a nation led by Marxist authoritarians who loathe American ideals and who are cozy with China, North Korea, and Russia. After converting to Christianity, Sami was put on an Iranian government hit list and escaped on foot to Europe with nothing more than the clothes on his back. He's never lived in America, but he believes the stability of the world is dependent upon the political future of the USA.

All shades of Marxism are evil. As for the United Nations? It represents the think tank responsible for spreading Marxist doctrine to the world. After all the death, misery, and destruction that socialism and communism have inflicted, it is truly mind-boggling when present-day nations choose the same evil path—like Russia (where Marxism still runs deep and political dissidents are imprisoned), China (where one million Uyghur Muslims and hundreds of thousands of Christians are in concentration camps), Iran (where Christians and women are horribly mistreated), Cuba (where hundreds remain in prison for participating in the 2021 freedom protests), Pakistan (another Marxist-growing hotbed, where Western journalists are abducted)—they're all given a seat on the UN Human Rights Council!

The United Nations' supreme assembly is the Security Council, a body where the first and foremost nuclear powers (the United States, Russia, China, France, and the United Kingdom) have a veto. Because these five powers rarely find themselves on the same side of a conflict, the UN hardly attempts to intervene in conflicts anywhere in the world.

For example, as of this writing, Russia-Ukraine conflict continues. The UN's failure to prevent one-fifth of its permanent Security Council from overrunning another member state is a lesson in organizational hypocrisy. At the onset of the Russian bombing in Ukraine, the best UN secretary-general António Guterres could utter was, "I have only one thing to say, from the bottom of my heart: President Putin, stop your troops from attacking Ukraine. Give peace a chance."

Stephen Daisley writes in the *Spectator*, "The UN Charter reads like self-satire and, while there's plenty of hypocrisy in international relations, it's never good when your organization's objectives double as punchlines."[1]

Regardless of their fecklessness and differing cultural backgrounds, representatives at the UN are generally unified by secular humanism, elitism, and the unyielding doctrine of climate change threatening the planet and sustainable development being the cure. Because so many member states are built on variations of Marxist philosophy, there is much agreement in principle that the inalienable rights baked into the fabric of America must be ripped out.

The UN is banking on sustainable development as their panacea. The UN's António Guterres said, "Humanitarian response, sustainable development, and sustaining peace are three sides of the same triangle."[2]

---

1   Stephen Daisley, "What Is the Point of the UN?," *Spectator*, February 24, 2022, https://www.spectator.co.uk/article/what-is-the-point-of-the-un/.

2   António Guterres, Speech to the United Nations General Assembly, December 12, 2016.

No, the triangulation Guterres is talking about is the UN's scheme to divide and conquer the world.

PRIVATE LAND BAN

A teacher's guide on sustainable development, produced by Vanderbilt University, reads:

> Sustainability offers a novel framework for asking enduring philosophical questions: What is the good life? How do we create a better world? Thinking and teaching about sustainability are future-oriented projects, but the relevance of sustainability principles and practices must be articulated in the present.[3]

A *good life*. A *better world*. Wow. Who wouldn't be for all that?

But the terms "good" and "better," as are all the terms used by the climate change cultists, are subject to the intended meaning applied by the instructor and reinforced by a steady dose of student indoctrination.

To fully grasp the roots of sustainable development, we must look back to an extremely significant environmental event in 1976, six years after the first Earth Day—the United Nations Conference on Human Settlements, or "Habitat I," in Vancouver, Canada.

Remember, the Left works very patiently, and the term "sustainable development" was not yet brought forward at this gathering.

---

3   Vanderbilt University, Center for Teaching, "Teaching Sustainability," https://cft. vanderbilt.edu/guides-sub-pages/teaching-sustainability/.

CLIMATE CULT

The subject of land was the focus of Habitat I, specifically private land ownership. Agenda Item 10 presented the UN's official policy regarding this key issue. The document's preamble could easily be confused with a draft written by Marx or Lenin:

> Land...cannot be treated as an ordinary asset, controlled by individuals and subject to the pressures and inefficiencies of the market. Private land ownership is also a principal instrument of accumulation and concentration of wealth and therefore contributes to social injustice; if unchecked, it may become a major obstacle in the planning and implementation of development schemes. The provision of decent dwellings and healthy conditions for the people can only be achieved if land is used in the interests of society as a whole. Public control of land use is therefore indispensable....[4]

To underscore:

- Private land ownership...contributes to social injustice.
- If unchecked, it may become a major obstacle in the planning and implementation of development schemes.
- Public control of land use is therefore indispensable.

I was recently in Bulgaria, visiting friends and speaking at a church in Sofia. Bulgarians are wonderful people, despite the fact that their country was decimated by Soviet occupation after World War II, which continued until the collapse of the Soviet

---

4   *United Nations Conference on Human Settlements*, Agenda Item 10, "Preamble," May 31–June 11, 1976.

Union in 1991. To this day, everywhere you look, there are large, ugly apartment buildings—stack-and-pack housing—built and formerly owned by the communist regime. The Soviets had a plan very much like what the UN agreed to in Habitat I, which also included a declaration stating that "adequate shelter and services are a basic human right."[5]

In 1944, when the Balkan region of Europe fell into the hands of Russia, all communist opposition was violently eliminated. Social life, culture, and the education system were turned into tools of communist propaganda. Unreliable teachers from universities and schools were purged. Religious education was banned; Russian language classes became obligatory. All publishing was controlled by the state. Censorship was introduced to rewrite history. Any contact with the free world was forbidden.

But among the highest priorities for the communists were land and personal property. Farms were seized by the state, estates were impounded, all private industry and commerce were taken over.

The communists boasted at the time, "The rent is free," as was electricity and heating. Never mind the fact that apartment buildings were in a constant state of disrepair, and the energy supplied by the government was sporadic at best, and that fur coats were often worn to bed during winter months to keep warm. Forget that the plumbing was usually backed up and elevators broken or nonexistent. All of Eastern Europe was a surveillance state, with life discouraged outside of the city. Movement and associations could be easily tracked, yet authoritarian planners continually lied that theirs was a better world and a good life.

---

5   Ibid., 7.

There was no social injustice; everyone was equal; there was little crime—or so they were told. Behavior was easily controlled. Everyone was on the same social footing—except for the elite leadership and their exclusive accomplices. The workers—the lesser-minded—were pacified with sports, television, and vodka.

It has been more than thirty years since Soviet communism miserably collapsed in Europe, but the stench of its creed still lingers.

That preamble from Habitat I may as well have been cut and pasted from the *Communist Manifesto*: "In this sense, the theory of the Communists may be summed up in the single sentence: Abolition of private property."

In their opinion, private land ownership contributes to social injustice and directly interferes with the UN's so-called development scheme.

And what a scheme it is.

OUR COMMON FUTURE

In 1987, a decade after revealing their unabashed goals for privately held land, the United Nations' plot was made public, although cloaked in the Orwellian-like expression, "sustainable development."

According to the Vanderbilt teacher's guide, "The term 'sustainability' has an important history in development literature."

Yes, it does—it promotes radical activism.

The phrase was first unveiled in the *Brundtland Report*, also referred to as *Our Common Future*, published by the UN's World Commission on Environment and Development. Vanderbilt praises *Our Common Future* as if it were manna from heaven.

In fact, almost every sustainable development teaching literature I've examined references *Our Common Future*.

Other cunning terms were brought forth in the UN report. Earth was renamed a "biosphere"; compelled transfer of wealth became "economic growth"; energy rationing dubbed "life-style adjustment."

Utilizing further mesmeric language, *Our Common Future* defines sustainable development:

- The concept of sustainable development does imply limits—not absolute limits but limitations imposed by the present state of technology and social organization on environmental resources and by the ability of the biosphere to absorb the effects of human activities…sustainable development requires meeting the basic needs of all and extending to all the opportunity to fulfill their aspirations for a better life.[6]

- Meeting essential needs requires not only a new era of economic growth for nations in which the majority are poor, but an assurance that those poor get their fair share.…[7]

- Sustainable global development requires that those who are more affluent adopt lifestyles within the planet's ecological means—in their use of energy, for example.[8]

Sustainable development is far more than recycling cans, bottles, and cardboard, eating tofu, or driving a Tesla—that's only its

---

6   *Our Common Future: Report of the World Commission on Environment and Development*, "Chairman's Foreword," subchapter 3.27, 1987.
7   Ibid., subchapter 3.28.
8   Ibid., subchapter 3.29.

feel-good public persona. Sustainable development is the practical application of communist theory with the environment as its hook. It is a reality that Marx and his early disciples could only dream of. Notice the rhapsodic attributes expressed in the plan:

- Meeting the basic needs of all...

- The opportunity to fulfill aspirations for a better life...

- A new era of economic growth for the nations...

- Those poor get their fair share...

- Adopt lifestyles within the planet's ecological means...

This is the kind of talk associated with charlatans, snake oil salesmen, and desperate politicians. But the document goes on to diabolically declare a surefire enemy, one that is actually invisible: carbon dioxide. *Our Common Future* officially introduced CO_2 as not only the cause of global warming but a threat to world peace and an agent of economic inequality:

- With the exception of CO_2, air pollutants can be removed from fossil fuel combustion processes at costs usually below the costs of damage caused by pollution. However, *the risks of global warming make heavy future reliance upon fossil fuels problematic.*[9]

- *All nations may suffer from the releases by industrialized countries of carbon dioxide* and of gases that react with the ozone layer, and from any future war fought with the nuclear arsenals controlled by those nations. All nations

9   Ibid., Chapter 7, subchapter 2.18.

will also have a role to play in securing peace, in changing trends, and in *righting an international economic system that increases rather than decreases inequality, that increases rather than decreases numbers of poor and hungry* [emphasis mine].[10]

The faceless UN bureaucrats who crafted *Our Common Future* were hopeful that their new prod would be their most effective instrument yet to bring about global societal and political change. The key would be getting the United States on board, because—just as in war, or even a bar fight—once the biggest guy goes down, a winner is born.

UNPACKING SOCIAL JUSTICE

With the rollout of the *Brundtland Report*, sustainable development became the UN's new ethos, commandment, and stratagem, masterfully tying the environment, CO_2, and global warming to the economics of something spoken of only in liberal academic circles at that time: "social justice."

The concept of social justice was first developed in 1843 by Jesuit philosopher Luigi Taparelli d'Azeglio. His term attempted to describe a type of virtue required for the success of postagrarian societies. D'Azeglio's primary question was: Does the state have the right to enforce personal virtue or coerce its citizens to behave in a certain way?

The United Nations responded with a resounding *yes!* in a 146-page 2006 paper, *Social Justice in an Open World*, a docu-

---

10   Ibid., Chapter 12.3.

ment rife with flowery utopian language that can be summed up in a couple of its sentences:

- Social justice is not possible without strong and coherent redistributive policies conceived and implemented by public agencies.[11]

- Social justice will only flourish if environmental preservation and sustainable development constitute an integral part of growth strategies now and in the future.[12]

First, social justice requires forceful and orderly redistribution of personal assets—finances and property—carried out by the government. Second, this coerced redistribution of assets will only progress by using the cause of environmental preservation as the basis for the "growth strategies."

This is classic communism. But what are growth strategies?

Look no further than what this same document puts forward in a section entitled "Social Justice: A Recent and Politically Charged Concept." Several philosophers are named in the paragraph, though one is purposefully left unidentified.

> The concept of social justice and its relevance and application within the present context require a more detailed explanation. As mentioned previously, the notion of social justice is relatively new. None of history's great philosophers—not Plato or Aristotle, or Confucius or Averroes, or even

---

11   The International Forum for Social Development, Department of Economic and Social Affairs, Division for Social Policy and Development, *Social Justice in an Open World: The Role of the United Nations* (New York: United Nations, 2006), 6.
12   Ibid., 7.

Rousseau or Kant—saw the need to consider justice or the redress of injustices from a social perspective.

Note the very next sentences from this same paragraph. They succinctly explain that the "concept" of modern social justice "first surfaced in Western thought and political language in the wake of the industrial revolution and parallel development of the socialist doctrine." The names of people who brought forth this "concept" are omitted—Marx and Engels:

> The concept first surfaced in Western thought and political language in the wake of the industrial revolution and the parallel development of the socialist doctrine. It emerged as an expression of protest against what was perceived as the capitalist exploitation of labour and as a focal point for the development of measures to improve the human condition.[13]

Social justice is "an expression of protest against" capitalism sanctioned by the United Nations. Like so many proposals coming from the UN, this paper makes clear that governments are charged with the responsibility to change human behavior through education, regulation, and laws. As practiced, this current version of social justice fosters economic egalitarianism through progressive taxation, income and property redistribution, discrimination, and censorship. These programs ultimately bring great harm to humanity, as the past 120 years have proven. In

---

13   Ibid., 11–12.

every location where socialism/communism has been attempted on a national scale, it has failed to remove class distinctions from society. Instead, it replaces various divisions with an intellectual elite class versus everyone else.

Now, if by practicing social justice it means that individual members of society have a moral obligation to care for those less fortunate, then that is correct. Indeed, Jesus said the greatest biblical commandment is, "You shall love your neighbor as yourself."[14] In other words, each human being is encouraged to do what he or she can to help the lowly. But today's politicized perception of social justice replaces individual charity with government bureaucracies and services that promote compulsion, intimidation, and an outright taking of that which was rightfully earned to fund welfare programs that generally prove wasteful and incompetent. The brand of social justice dictated today does not inspire personal human interaction and heartfelt giving but instead fosters resentment and further division.

"WRENCHING TRANSFORMATION"

In 1992, five years after the introduction of sustainable development, the United Nations held their inaugural "Earth Summit" in Rio de Janeiro, Brazil. Virtually every country on the planet was represented, with 108 heads of state in attendance—including politically struggling President George H. W. Bush. Some ten thousand journalists were there to witness the unveiling of the summit's centerpiece: *Agenda 21*.

---

14 Matthew 22:39 (New International Version).

It was a clarion call towards "a global partnership for sustainable development."[15]

Whereas the *Brundtland Report* introduced the world to new terminology and provided a diagnosis of the planet's supposed ills, *Agenda 21* details envisioned cures. The sweeping scheme articulates a plan to end poverty, formally declares adequate housing as a "right,"[16] and urges each country to provide its citizens with universal health care.[17]

Some might ask, "And what's wrong with that?"

What's wrong is that it is all brought about through involuntary institutions of social justice.

Agenda 21 is built upon concepts brought forth in *Our Common Future*, thus, cementing a relationship between sustainable development, the environment, and social justice. The nations of the world were empowered to accomplish the following:

- Create enforceable "environmental laws and regulations that are based upon sound social, ecological, economic, and scientific principles."[18]

- Demand "the need to control atmospheric emissions of greenhouse and other gases and substances,"[19] as well as "promote appropriate energy efficiency and emission standards."[20]

---

15 *Agenda 21*, United Nations Sustainable Development, "Preamble," June 1992, https://sdgs.un.org/sites/default/files/publications/Agenda21.pdf.
16 Ibid., Chapter 7.6.
17 Ibid., Chapter 6:13.
18 Ibid., Chapter 8.14.
19 Ibid., Chapter 9.9.
20 Ibid., Chapter 9.12 j.

- Schools are to be utilized to "encourage education and awareness-raising programs at the local, national, subregional, and regional levels concerning energy efficiency."[21]

- "Population policy should also recognize the role played by human beings in environmental and development concerns."[22]

- Abortion is the recommended population control method of choice, i.e., "curative health facilities...for the responsible planning of family size."[23]

Agenda 21 was signed by George H. W. Bush. Although legislation was never passed by Congress to mandate the agenda, it really did not matter. Five months after Rio, one-term Bush lost the White House to President Bill Clinton and Vice President Al Gore, two cunning politicians who would carry the sustainable development baton more adeptly than Bush ever could. According to Gore, sustainable development would achieve a "sacrifice, struggle, and a wrenching transformation of society."[24]

PLUS, SOCIAL EQUITY

The term "social equity" is credited to H. George Frederickson from the University of Kansas, who began using the idiom in 1968 to describe social equity as a pillar of public bureaucracy.[25]

---

21 Ibid., Chapter 9.12 k.
22 Ibid., Chapter 5, 5.3.
23 Ibid., Chapter 5, 5.51.
24 Al Gore, *Earth in the Balance*, (paperback version, Plume, 1993), 274.
25 H. George Frederickson, "Public Administration and Social Equity," *Public Administration Review*, vol. 50, no. 2 (March–April 1990), 228–237.

The UN capitalized on Frederickson's phrase and expanded its definition.

Social equity has become the expected outcome of sustainable development in this way: sustainable development brings forth social justice, which leads to social equity. And just to be clear, social equity has nothing to do with social equality.

Illustrating the point, *Our Common Future* declares:

> [Sustainable] development involves a progressive transformation of economy and society... Even the narrow notion of physical sustainability implies a concern for social equity....[26]

The progressive transformation of the economy and society implies a complete reset to enact previous goals stated earlier, namely the abolition of private land ownership and the redistribution of wealth. The aforementioned 2006 UN paper, *Social Justice in an Open World*, reveals social equity is anything but just:

> Equity further demands that international cooperation be built upon processes, rules, and agreements that give preference and advantages to the weakest, and not based on the "equal" treatment of partners that are, objectively speaking, very unequal.[27]

---

26 *Our Common Future: Report of the World Commission on Environment and Development,* United Nations, March 20, 1987, Chapter 2, IV.3.

27 The International Forum for Social Development, Department of Economic and Social Affairs, Division for Social Policy and Development, *Social Justice in an Open World: The Role of the United Nations* (New York: United Nations, 2006), 141.

Presented in a simpler way, sustainable development is the environmental agenda's business plan, while social equity is the anticipated result.

In fact, Maurice Strong, founder of the United Nations Environment Program and architect of *Our Common Future*, described sustainable development to the *Canadian Business Review* in a 1990 interview: "It's like putting our planet, Earth Incorporated, if you will, on a sound business basis."[28]

Revealing the breadth of his plan, Strong explained, "Now, sustainable development is a broad concept and it must be defined in virtually every sector of human activity…Development requires economic growth but it requires the kind of qualitative economic growth that permits us to meet social and human needs, which, of course, are the purposes of growth in the first place."

Strong's comments align with Marx's laws of matter, endorsing an elitist mentality and allowing some to perceive themselves as god, capable of meeting the needs of humanity through the inventions of their self-proclaimed "enlightened" minds.

The economic goals of the climate cult require authoritarian overlords who will dole out resources and pick winners and losers, demanding submission, and employing aggression, all as they, alone, see fit. In the case of equity and social justice, revenge for past grievances will be exacted at the behest of government laws, regulations, and pressures stemming from concepts like ESG.

Recall, liberty, as enshrined in the founding documents of the United States, was to protect *We the People* from such madness. A response to today's woke version of equity could have come from a sarcastic statement made by Massachusetts sena-

---

28 Catherine Johnson, "Our Common Future," *Canadian Business Review*, March 22, 1990.

tor Daniel Webster in 1830, who noted, "The constitution was made to guard people against the nature of good intentions."[29]

The "good intentions" of the UN and their co-conspirators created a one-two-three punch—sustainable development, social justice, and social equity, the antithesis of life, liberty, and the pursuit of happiness.

IN THEIR OWN WORDS

The following quotes are from a post at the Democratic Socialists of America (DSA) website entitled "Capitalism, Socialism and Sustainability." It confirms apprehensions you may have about the intentions of the world's power brokers when it comes to the buzzwords "sustainability," "social justice," and "social equity" (emphasis within each point is mine):

- Securing an environmentally sustainable production system *will require fundamental political and social change on every scale* from household to planet.

- Human and environmental needs can be brought into sustainable balance only if production takes account of all environmental consequences. *This requires conscious planning and foresight. But who will do the planning?*

- A sustainable economy requires a system in which production is democratically planned and controlled by well-informed people. The environment can be sustained by collective stewardship as *our material needs are securely*

---

29 James Rees, *The Beauties of the Hon. Daniel Webster; Selected and Arranged, with a Critical Essay on His Genius and Writings* (1839), 30.

met by a fair distribution and sharing of resources, and our psychological needs are met through an ethos fostering coop-eration rather than acquisition and competition. We call such a system democratic socialism.

- *A socialist society can be achieved by nonviolent struggles, continuing over generations,* to expand democratic rights, institutions, and social relations within a mainly capital-ist system, until the democratic processes and structures come to predominate.

CONCLUSION

Reading those four bullet points from the DSA should cause us to seriously ponder what kind of a world are our kids and grand-kids going to live in.

How will *fundamental political and social change on every scale* be made manifest in every *household* on the planet?

I'm guessing they don't expect this to be on a voluntary basis.

The DSA asks the question, *who will do the planning?*

Unelected bureaucrats, lying politicians, authoritarian dictators?

And, if *a socialist society can be achieved by nonviolent strug-gles, continuing over generations,* are we to forget all the violence and bloodshed that communism caused in the twentieth cen-tury with Marxist takeovers in Russia, Eastern Europe, China, and North Korea? And add to that list Laos, Vietnam, Cuba, Venezuela, and the many African nations that have fallen into a socialist/communist stupor.

And remember, the climate cult rejects history. As Marx said, "History does nothing; it possesses no immense wealth; it wages no battles."[30]

Marx understood that those controlling the means of production have the power to control the culture. Therefore, by vilifying the invisible by-product of industrial production—carbon dioxide—as a pollutant, the elites plan to become masters of the universe. Banning drilling for oil, outlawing the sale of vehicles with combustion engines, and eliminating gas stoves is just scraping the surface of their sinister plans. This is about authoritarians controlling humanity—you and me. Unless common sense prevails, the energy grid will *never* support the climate agenda's master plan. And that's their intention.

Spurious dreams of "a good life" and "a better world" are distinctly linked to the thorough concept of sustainable development. While cult planners envision all the nations of the world coming together to accomplish these plans, the nation most radically impacted will be the United States of America.

Sustainable development and its branches of social justice and social equity will only be able to take hold if the inalienable rights of life, liberty, and the pursuit of happiness achieved through personal property ownership are smashed and trampled into the dirt. Propaganda, indoctrination, and fear are the globalists' recipes to expand the cult and recast the world in their image. "Climate change is here. It is terrifying. And it is just the beginning," the UN's Guterres recently claimed, warning that the tragic consequences will be "children swept away by monsoon rains, families running from the flames, workers collapsing

---

30   Marx, *The Holy Family*, Chapter Six (1845).

in scorching heat. The era of global warming has ended; the era of global boiling has arrived."[31]

Many are of the opinion that the United Nations has existed as just another bureaucratic, money-stealing monolith that really has never accomplished anything beneficial to mankind. While that assessment is fairly accurate, we mustn't underestimate the real and growing threat that they pose to humanity.

31 "Hottest July Ever Signals 'Era of Global Boiling Has Arrived' Says UN Chief," UN News, July 27, 2023, https://news.un.org/en/story/2023/07/1139162.

CHAPTER SEVEN

RESET

THE VAGRANT AMBASSADORS AND BUMBLING bureaucrats at the United Nations, known for endless empty decrees and mountains of paperwork, teamed up with their rich and powerful friends at the World Economic Forum to produce a press release back in 2019:

> **New York, USA, 13 June 2019**—The World Economic Forum and the United Nations signed today a Strategic Partnership Framework outlining areas of cooperation to deepen institutional engagement and jointly accelerate the implementation of the 2030 Agenda for Sustainable Development. The framework was drafted based on a mapping of existing collaboration between the two institutions and will enable a more strategic and coordinated approach towards delivering impact.
>
> The UN-Forum Partnership was signed in a meeting held at United Nations headquarters between UN Secretary-General António Guterres

and World Economic Founder and Executive Chairman Klaus Schwab.

"Meeting the Sustainable Development Goals is essential for the future of humanity. The World Economic Forum is committed to supporting this effort, and working with the United Nations to build a more prosperous and equitable future," said Klaus Schwab, World Economic Founder and Executive Chairman.

"The new Strategic Partnership Framework between the United Nations and the World Economic Forum has great potential to advance our efforts on key global challenges and opportunities, from climate change, health, and education, to gender equality, digital cooperation, and financing for sustainable development. Rooted in UN norms and values, the Framework underscores the invaluable role of the private sector in this work—and points the way toward action to generate shared prosperity on a healthy planet while leaving no one behind," said António Guterres, UN Secretary-General.[1]

With this partnership, the UN essentially became "United Nations, Inc." The agreement represents the corporate capture of the UN. Although corporate influence has long been wielded

---

1   News Release, "World Economic Forum and UN Sign Strategic Partnership Framework," World Economic Forum, June 13, 2019, https://www. weforum.org/press/2019/06/world-economic-forum-and-un-sign-strategic-partnership-framework/.

within the United Nations' system, under the new terms between the UN and WEF, the world's wealthy elites—the Davos crowd—have become even more influential, prosperous, and dominant in this corrupt system.

As might be expected, this arrangement had been in the works for many years.

In 2009, the WEF published an extensive, six-hundred-page report titled *The Global Redesign Initiative* (GDI). This epistle called for a new system of global governance—a new world order—one in which the decisions of state governments would be secondary to multistakeholder-led initiatives launched by the WEF, allowing corporations and their wealthy elite majority owners a defining role in running the planet.

The ultimate winners would be the highly favored members of the Davos crowd and their stakeholders.

The losers? Everyone else on the planet.

The 2009 *Global Redesign Initiative* telegraphed the WEF's intention to partner with the UN, putting Schwab's much-heralded plan for a "Great Reset" into motion.

STAKEHOLDER THEORY

Despite a great deal of political, media, and educational misinformation, the Great Reset is *not* a conspiracy theory created by right-wingers. It is a stakeholder theory created by founder Schwab, who, like Marx, believes that ultimate power belongs to those with the highest knowledge and intellect. The WEF has spelled out their many conceptual intentions via endless papers and articles on their website. And, in large measure, their schemes rely on utilizing climate change as a lever to squelch personal liberty, greatly modify property rights, reduce national sovereignty, and treat the bulk of humanity as a class of common workers and welfare recipients.

Marx would be dancing a jig.

As executive chairman of the World Economic Forum, Schwab suggests the Great Reset is merely an attempt to address the weaknesses of capitalism, as well as the looming catastrophes posed by anthropogenic climate change and other associated environmental deterioration.

In 1971, Schwab, then an engineer and economist by training, created the European Management Forum, later renamed the World Economic Forum. That same year, Schwab published his first book, *Modern Enterprise Management in Mechanical Engineering*, in which he introduced the concept of "stakeholder capitalism." The term is defined in a PowerPoint presentation on the WEF website from the 2021 Davos event as "a form of capitalism in which companies do not only optimize short-term profits for shareholders, but seek long-term value creation by taking into account the needs of all their stakeholders and society at large."[2]

Thus, stakeholder theory is the same nuts and bolts of ESG originally conceptualized by the UN.

Activists are constantly bombarding directors and majority stockholders of large-scale corporations with demands to be socially responsible in the name of the environment, social justice, and social equity. Knowing these agitators are relentless and quite capable of jolting the bottom line via lawsuits, protests, and

---

2   Klaus Schwab, Peter Vanham, "Stakeholder Capitalism," *The Davos Agenda*, January 22, 2021.

boycotts, such requests are met by managers who are only too happy to satisfy the extortioners, either through amplified lip service or by deed. By acknowledging the complaints and concerns of the protestors, the more prominent managers gain a seat at the meeting room table as "stakeholders."

A LinkedIn article actually provides pointers on how to pacify stakeholders:

> Managing stakeholder expectations and relationships is a key skill for project managers and leaders. One way to enhance stakeholder engagement and satisfaction is to design and implement a stakeholder recognition and reward system. This system can acknowledge the contributions, achievements, and feedback of stakeholders, and motivate them to continue supporting the project or organization.[3]

Handling stakeholders is much like mommy mollifying her kid who has gone into a candy tantrum in the checkout line. Adding these participants to the corporate structure makes for warm and fuzzy feelings within the company and wards off nasty attacks from activists that could create bad press that impacts shareholder value. The glossy annual reports of many publicly traded companies are prefaced by soothing references to what they have done for their organization's many stakeholders.

---

3   AI-powered collaborative article, "What Are Some Best Practices for Designing and Implementing a Stakeholder Recognition and Reward System?," LinkedIn, https://www.linkedin.com/advice/0/what-some-best-practices-designing-implementing-1c.

In reality, stakeholder theory undermines the defining feature of capitalism: the exclusive rights of ownership. Likewise, it sabotages the inalienable right found in property ownership. The theory also allows those with no real skin in the game to call the shots, getting what they demand, and simultaneously profiting. It's like the screaming child in the checkout line walking out of the store with not just a candy bar but also a bag of cheese puffs and a big grin.

*"PLAN*DEMIC?*"*

The phrase "Great Reset" originally appeared in a 2010 book by social theorist, Richard Florida, *The Great Reset: How New Ways of Living and Working Drive Post-Crash Prosperity*, a call-to-action work following the 2008 financial crisis. Klaus Schwab apparently appropriated the phrase to represent his stakeholder vision for a new kind of capitalism and world system.

At the 2014 annual WEF Davos meeting, Schwab declared:

> We need to push the reset button. The world is still much too much caught in a crisis-management mode. We should look at our future in a much more constructive and strategic way. That is what Davos is about…There is no place in the world where so many stakeholders of our global future assemble, all united by the mission of improving the state of the world.[4]

---

4   Kim Hjelmgaard, "Push 'Reset' Button on World, WEF Founder Says," *USA Today*, January 15, 2014.

In 2017, the WEF published its reset plan: "We Need to Reset the Global Operating System to Achieve the SDGs [Sustainable Development Goals]. Here's How."[5]

Next, the WEF organized two major events that eerily seemed to anticipate the COVID-19 outbreak. In May 2018, the WEF collaborated with the Johns Hopkins Center for Health Security to conduct the CLADE X exercise, or what they referred to as a "tabletop" simulation of a national response to a massive pandemic.

According to one report, the CLADE X simulation demonstrated "the lack of both a protective vaccine and a proactive worldwide plan for tackling the spread of a catastrophic global pandemic resulted in the death of 150 million people across the Earth."[6]

Less than a half-year later, in October 2019, the WEF teamed up with the Bill and Melinda Gates Foundation at Johns Hopkins to stage another pandemic exercise, Event 201. The gathering simulated an international response to the outbreak of a novel coronavirus, two months before the COVID-19 outbreak became international news.

CLADE X and Event 201 simulations anticipated practically every eventuality of the COVID crisis, including responses by governments, health agencies, conventional media, social media,

5   Homi Kharas, "We Need to Reset the Global Operating System to Achieve the SDGs. Here's How," World Economic Forum, January 13, 2017, https://www.weforum.org/agenda/2017/01/we-need-to-upgrade-the-sustainable-development-goals-here-s-how/.

6   Kim Riley, "Mock Clade X Pandemic Decimates Human Population; Denotes Global Pre-Planning Needs," Homeland Preparedness News, May 21, 2018, https://homelandprepnews.com/countermeasures/28548-mock-clade-x-pandemic-decimates-human-population-denotes-global-pre-planning-needs/.

and elements of the public. These procedures caused many to wonder, *was this a pandemic or a plan-demic?* Could it be the COVID-19 catastrophe was staged by the actors associated with the WEF as an alibi for initiating the Great Reset? An honest bystander might find the timing of these events and the outbreak of the virus extremely curious.

Then came several key moves.

First, the establishment of the WEF's formal partnership with the United Nations in June 2019. A year later, the WEF held its Great Reset Summit at the fiftieth annual meeting of the World Economic Forum in June 2020. Attendees included a host of global governance advocates, including the UN secretary-general, the president of the European Central Bank, the secretary-general of the Organization for Economic Co-operation and Development, the managing director of the International Monetary Fund, union leaders from select countries, and representatives of progressive political and environmental organizations. England's Prince Charles was also in attendance, saying, "The threat of climate change has been more gradual" than COVID-19 and "its ever-greater potential to disrupt, surpasses even that of COVID-19."

The man who has since become Britain's king continued, "We have a golden opportunity to seize something good from this crisis—its unprecedented shockwaves may well make people more receptive to big visions of change." Adding later, "It is an opportunity we have never had before and may never have again."[7]

---

7   Justin Haskins, "Introducing the 'Great Reset,' World Leaders' Radical Plan to Transform the Economy," *The Hill*, June 25, 2020, https://thehill.com/opinion/energy-environment/504499-introducing-the-great-reset-world-leaders-radical-plan-to/.

In 2021, the WEF rolled out details of the reset. Former senator John Kerry, the Biden administration's special presidential envoy for climate, addressed Davos with extensive US plans for the reset:

> President Biden will sign another series of executive orders that continue to advance his climate agenda. First, making climate central to foreign policy planning and national security preparedness by creating platforms to coordinate climate action across all federal agencies and departments, by directing his administration to develop a U.S. climate finance plan, as well as a plan for ending international financing of fossil fuel projects with public money, and moving to ratify the Kigali Amendment to the Montreal Protocol, and by hosting a leaders summit less than three months from now on Earth Day, April 22.[8]

Kerry's remarks reveal a sharp turn from the way US foreign policy historically has worked to protect the nation's political, cultural, and economic influence throughout the world. The Democrats' foreign policy, including national security, is centered on climate—complete with the development of "climate finance plans" and the defunding of energy projects involving proven resources, namely, fossil fuels.

And Kerry wasn't just all talk. In 2023, two days before Earth Day, President Biden announced the United States was provid-

8   John Kerry, "Remarks at World Economic Forum, Davos 2021," US Department of State, January 27, 2021, https://www.state.gov/remarks-at-world-economic-forum-davos-2021/.

ing $1 billion to the UN's Green Climate Fund, bringing total US contributions to $2 billion. This is in addition to his 2021 pledge to quadruple US climate support for developing countries to more than $11 billion a year by 2024.[9]

Research by an anti-corruption whistleblower organization reveals what a huge waste of taxpayer money these giveaways are. Corruption in climate finance is rampant, with the top recipient nations being "the riskiest places in the world for corruption" and yet receiving "41.9% of all climate-related overseas development assistance."[10]

QUESTION WHAT IT MEANS TO BE HUMAN

In a 2021 article promoting his bestselling book, *COVID 19: The Great Reset*, Schwab proposed that that year, 2021, be designated as "Year Zero," a term originally coined immediately after the end of World War II and signifying the rebuilding of the world's war-torn regions. Applying superfluous tones that one might expect from a presidential candidate, Schwab wrote, "This time, the focus is on the material world but also on so much more. We must aim for a higher degree of societal sophistication and create a sound basis for the well-being of all people and the planet." He continued clearly taking aim at the US, insisting that free markets and limited government constitute a model that has "broken down," emphasizing the need for a Great Reset.

---

9   The White House Briefing Room, "FACT SHEET: President Biden to Catalyze Global Climate Action through the Major Economies Forum on Energy and Climate," April 20, 2023.

10 Michael Nest, Saul Mullard, Cecilie Wathne, "Corruption and Climate Finance," CMI, U4Brief 2020:14, https://www.u4.no/publications/corruption-and-climate-finance.pdf.

In his book, Schwab explains the COVID crisis should be regarded as an "opportunity...to make the kind of institutional changes and policy choices that will put economies on the path toward a fairer, greener future."[11]

It may be challenging to connect the dots, but it's important to try. Sustainable development is the climate agenda's business plan; ESG is its practical application. COVID provided a pretext for enactment as the cult incredibly and easily fell in line with mandates and protocols, essentially producing a new normal throughout the developed world. A rollout of a "greener future," designed to expand stakeholder influence, produce stagflation, and force lifestyle adjustments for the working class, provided more handouts for additional welfare recipients, ramped up the CO_2 emissions fixation, and began moves within financial markets that helped the elite class to prosper. The new normal necessitated the confluence and transformation of economic, technological, medical, genomic, environmental, and governance systems—a total reset, an alteration of the world as we know it, one that will go so far as to cause us, in the words of Schwab's WEF, "to rethink...even what it means to be human."[12]

Michael Rectenwald, a foremost scholar on the topic of secularism, provides a succinct forecast of WEF intentions:

> [T]he Great Reset would involve a consolidation of wealth, on the one hand, and the likely issuance of universal basic income (UBI) on the

11 Klaus Schwab, Thierry Malleret, *COVID-19: The Great Reset* (Forum Publishing, 2020), 57.
12 "Fourth Industrial Revolution," World Economic Forum, https://www.weforum.org/focus/fourth-industrial-revolution/.

other. It might include a shift to a digital currency, including a consolidated centralization of banking and bank accounts, immediate real-time taxation, negative interest rates, and centralized surveillance and control over spending and debt.[13]

This is reminiscent of Al Gore's failed prediction that sustainable development would achieve a "sacrifice, struggle, and a wrenching transformation of society."

It's also the epitome of "Rahm's Rule."

Rahm Emanuel, then chief of staff to President-elect Barack Obama, famously stated, "You never want a serious crisis to go to waste. And what I mean by that [is] it's an opportunity to do things that you think you could not do before."

Emanuel believed that the 2008 fiscal crisis afforded the Obama-Biden administration the opportunity to "do things" they could not have otherwise. Trillions of dollars in higher spending later, US citizens are now faced with unimaginable national debt—over $33 trillion dollars to be exact at the time of this writing.

And then there is the $300 trillion that governments, households, and corporations around the world owe as estimated by the Institute of International Finance.[14] That figure is about

---

13 Michael Rectenwald, "What Is the Great Reset? Part I: Reduced Expectations and Bio-Techno-Feudalism," *Mises Wire*, December 16, 2020, https://mises.org/wire/what-great-reset-part-i-reduced-expectations-and-bio-techno-feudalism#footnote4_w1jx3np.

14 Nicole Goodkind, "The World Has a Major Debt Problem. Is a Reset Coming?," CNN Business, January 17, 2023, https://www.cnn.com/2023/01/17/investing/premarket-stocks-trading/index.html.

349 percent of global gross domestic product, the equivalent of $37,500 of debt for every single person in the world.

That is the real green crisis.

"We have at least four crises, which are interwoven," German vice chancellor Robert Habeck said at a WEF gathering of business leaders in 2022. "We have high inflation…we have an energy crisis…we have food poverty, and we have a climate crisis. And we can't solve the problems if we concentrate on only one of the crises."

Breaking down Habeck's list, the energy crisis is self-imposed—inexpensive and highly abundant fossil fuels are being demonized. From a scientific standpoint, the climate crisis is a sham; its hypothesis has shown itself to be vulnerable, and the climate forecasting models are imprecise. Global food poverty is the result of corrupt and tyrannical governance in poorer nations and, in more developed countries, the result of deep-pocketed advertising campaigns that urge people to consume unhealthful foods.

And inflation? The current surge has been driven, in large measure, by supply chain issues, increased fuel and energy costs, pent-up consumer demand, government economic COVID stimulus, and high interest rates imposed by the Federal Reserve. Paychecks, pensions, and retirement funds are being stretched beyond measure, often for too many to the breaking point. Many prominent money managers, like Andrew McCaffery, global CIO at Fidelity International, says these are all signatures of the reset: "Inflationary pressures have intensified, and supply chains are being redrawn. We see this as the start of the Great Reset."[15]

---

15   Brandon Russell, "The Start of 'The Great Reset'—Fidelity International's Q3 Outlook," *IFA Magazine*, July 13, 2022.

PREPPING

Many people keep a go-bag full of important items in the event that suddenly they have to flee their home from a fire, flood, or other emergency. Some keep a ready supply of nonperishable food and water on hand as well as ample cash, gold, and silver. Lots of people in the US are also very well-armed and trained in using their weapons. Basements have been secured for bunkers, if needed.

The high and mighty of this world are also prepping, though, but not for natural disasters. They're readying for Schwab's Great Reset, making loads of money now, because it may not be as easy in the near term. A venture capitalist friend recently told me, "I'm making as much money as fast as I can in the likely event the bottom of this economy falls out."

And the billionaire investors are doing the same. Case in point: during the first six months of the COVID outbreak, the wealth of the world's nearly 2,200 billionaires increased by 27.5 percent.[16] American billionaires saw their wealth grow by 62 percent between March 2020 and March 2022.[17]

While the world's wealthy make their rainy-day plans, those on the lower end of the pecking order face the imposition of universal basic income (UBI) championed by Schwab's Davos friends, progressive politicians, social justice and equity activists

16 Corneliu Pivariu, "Davos 2023: World Economic Forum, from Great Reset to Great Frangmentation," *Atalayar*, February 1, 2023, https://atalayar.com/en/blog/davos-2023-world-economic-forum-great-reset-great-fragmentation.

17 Juliana Kaplan, Madison Hoff, "Billionaires Saw Their Wealth Grow by 62% during the Pandemic, while Workers' Wages Grew by 10%," *Business Insider*, April 18, 2022, https://www.businessinsider.com/billionaires-wealth-grew-62-pandemic-while-workers-saw-10-raise-2022-4.

(stakeholders), and wealthy execs in Silicon Valley. And their message resonates. In 2011, a Rasmussen Poll found only 11 percent of Americans support UBI; that figure grew to 40 percent in 2020.[18] Proponents claim this free cash would break the cycle of dependency among the so-called disadvantaged, giving them time and money at last to seek the training and higher education they need to climb the economic ladder. Some even claim that UBI would help bring about a cultural revolution that would allow people free time to "live a dignified human life."[19]

Actually, this UBI insanity would lead to a society of laziness, sin, vice, and depression. It would rob individuals of liberty and the pursuit of happiness. A substantial portion of welfare recipients in the US are now generational, held captive by a system that was designed to help them. Free money might sound appealing but not when it runs out. Historically, each time coercive redistributive systems have been attempted, they have ended horribly, most notably in the former Soviet Union from 1922–1991. As I mentioned earlier, people in those former communist countries are still digging out all these many years later from the damage done by Lenin's, Stalin's, and Marx's madness.

The elite planners, however, do not see it the same way.

18 "Support Grows for Government-Run Health Care, Universal Basic Income," Rasmussen Reports, April 3, 2020, https://www.rasmussenreports.com/public_content/politics/current_events/healthcare/support_grows_for_government_run_health_care_universal_basic_income.

19 Doug MacKay, associate professor, University of North Carolina, quoted in "The Pros and Cons of Universal Income," *The Well*, March 10, 2021, https://college.unc.edu/2021/03/universal-basic-income/.

THE BRAVE NEW WORLD

The World Economic Forum issued a plan for the future of farming: 100 Million Farmers. According to the WEF, this agenda plans on "accelerating the transition towards food systems that are net-zero, nature-positive, and that increase farmer resilience. All three objectives are critical to achieve a successful transition."

Notice those "critical" objectives don't seem to imply putting more food on anyone's table. In fact, the WEF drills down on these objectives, explaining that they will:

- "Position food and water systems as a necessary pillar to meet food security, climate and biodiversity targets."

- "Accelerate action by working with farmers and growers, to drive action and scale adoption of climate-smart and regenerative practices."

- "Reach a systems tipping point of 100 million farmers and a billion consumers by unlocking transformative finance, data innovation and policy change."[20]

Students of history should find the gobbledygook from the WEF's plan chilling. It is reminiscent of the "Great Leap Forward," an economic plan executed by Mao Zedong and the Chinese Communist Party in 1958 and abandoned in 1961. The goal was to modernize the country's agricultural sector using communist economic ideologies. However, instead of stimulating China's economy, the Great Leap Forward resulted in mass starvation and subsequent chaos. Between thirty and forty-five

---

20 "100 Million Farmers," World Economic Forum, https://initiatives.weforum.org/100-million-farmers/home.

million Chinese citizens perished due to famine, execution, and forced labor—along with massive economic turmoil.

A curious mind has to contemplate, is this the WEF's plan as well? After all, progressives are united in agreement that the earth's current population of eight billion is unsustainable and must be reduced to less than two billion.

I have debated quite a few proponents of depopulation over the years, and I eventually ask each person the same question: How will such a massive reduction in population occur?

The more discriminating fall back to talking points taken from a paper edited by overpopulation guru Paul Ehrlich:

> There have been repeated calls for rapid action to reduce the world population humanely over the coming decades to centuries, with lay proponents complaining that sustainability advocates ignore the "elephant in the room" of human overpopulation. Amoral wars and global pandemics aside, the only humane way to reduce the size of the human population is to encourage lower per capita fertility.[21]

Amoral wars and global pandemics aside?

Does that not imply the tragedies of war and illness would be acceptable to bring forth the goal of a greatly reduced population? Is this really how they foresee the planet getting to the

---

21   Corey J. A. Bradshaw, Barry W. Brook, edited by Paul Ehrlich, "Human Population Reduction Is Not a Quick Fix for Environmental Problems," *PNAS*, vol. 111, no. 46 (October 27, 2014), https://www.pnas.org/doi/full/10.1073/pnas.1410465111.

optimum population that Ehrlich has spoken of—less than two billion people?[22]

Creepy. But it doesn't stop there.

TRANSHUMANISM

The original Industrial Revolution gave rise to factories powered by fossil fuels that changed the course of production, how people worked, and how they lived. The centralization of workplaces saw growing urbanization, an overall improved quality of life, and presented an opportunity for many to emerge from poverty.

However, the progressive argument believes that the Industrial Revolution deepened class division, pillaged the planet's resources, and loaded the atmosphere with carbon dioxide, which has spun the climate to a near out-of-control situation.

To counter, the WEF has introduced the "Fourth Industrial Revolution," which Schwab and his associates say will do away with the redundancy of human labor, replacing it with data and digital technologies.

Thus, "data is the new oil," as many in the Silicon Valley are fond of saying. Many firms are working on data and digital technologies that will restructure the labor market, education, communication, and life at large.

The Fourth Industrial Revolution is also often referred to as "transhumanism." This branch of science includes nanotechnology, genomics, robotics, and artificial intelligence (AI). There is

---

22 "The world's optimum population is less than two billion people," is Paul Ehrlich's exact quote from an interview with Damian Carrington, "Paul Ehrlich: Collapse of Civilisation Is a Near Certainty within Decades," *Guardian*, March 22, 2018, https://www.theguardian.com/cities/2018/mar/22/collapse-civilisation-near-certain-decades-population-bomb-paul-ehrlich.

even talk of penetrating human bodies and brains with technology, such as implanting microchips capable of reading and reporting on genetic makeup and the state of one's mental functions.

To many, this all seems wonderful.

For example, Elon Musk detailed his plans for a new brain-computer interface platform. American football star Tom Brady revealed his "brain resiliency program" designed to increase athletic edge. Elite US military forces are currently using brain stimulation headsets to boost mental and physical performance. Hundreds of firms and research labs are researching and developing new ways to help the human brain perform at higher levels for longer periods of time.

But the true masters of this new universe perceive something far more radical. For example, Julian Savulescu, director of the Oxford Center for Neuroethics, would like to accompany our physical enhancements with an improved system of morality via genetic engineering and hormone therapy to make us more cooperative and altruistic. In other words, we would all become like designer babies complete with a new personality and a fresh set of morals. Savulescu argues that since we already allow embryo selection and selective abortions to eliminate diseased embryos and fetuses, there should be no objection to using these methods to choose other genetic traits.[23]

If all of this sounds weird, and it should, we still need to be awakened to where this is headed. The world's largest transhumanist organization, Humanity Plus, states in its Manifesto:

---

23   Richard Weikart, "Can We Make Ourselves More Moral? Designer Babies, Hormone Therapy, and the New Eugenics of Transhumanism," June 6, 2016, https://lifeissues.net/writers/wei/wei_02makeourselvesmoral.html.

Transhumanism calls upon a heightened sensibility to reveal the multiplicity of realms yet to be discovered, yet to be realized. We are exploring how current and future technologies affect our senses, our cognition, and our lives...

Transhumanists invent and design with technology and collaborate with the cosmos, perform in multiple realities, automorph mind and body, conceive, innovate, and explore. We indelibly engrave longevity memes. We are the neo-cyberneticists utilizing high-end creativity, engineering skills, scientific data, and automated tools to author our visions.[24]

At the very heart of Humanity Plus's pie in the sky manifesto is their environmental platform, conceived by their executive director, Natasha Vita-More. In 1992, she was elected to a Council position in Los Angeles County, California, as a Green Party candidate who endorsed technology for mitigating environmental issues such as climate change. As revealed on her organization's website, transhumanism is another tangent of the climate change agenda:

The Transhumanist Platform was established to bring to the mainstream an awareness of emerging technologies that could be used to counter environmental problems, including pollution and climate change, and to protect all life forms

---

24 "The Transhumanist Manifesto," Humanity Plus, https://www.humanityplus. org/the-transhumanist-manifesto.

in a defensible, healthy ecosystem. However, this just one of the many issues that diplomatic policy-making needs to be realized.[25]

CONCLUSION

By now it should be clear. The climate agenda is central to a grand social engineering scheme to remake society into something perhaps best envisioned in Aldous Huxley's 1932 dystopian novel, *Brave New World*, wherein Huxley describes life under complete domination by totalitarian autocrats. Citizens are kept in a state of zombie-like, perpetual bliss, with every aspect of their lives controlled by powerful autocrats. Information and methods of communication are meticulously managed. Endless and mindless distractions prevent people from thinking, reasoning, or imagining. When stress or anxiety rises, drugs are dispensed. Morality is dead, and sexual promiscuity is encouraged. Minds have been twisted to delight in their own enslavement.

But until Huxley's vision is fully realized, scientists working in association with the United Nations and the World Economic Forum are racing forward with plans to spare the world from a temperature reading they contend is the maximum we collectively can withstand: 1.5 degrees Celsius.

Since mankind began utilizing fossil fuels at the dawn of the Industrial Revolution, the temperature has been steadily rising. By obliterating the previously discussed Medieval Warm Period, ignoring the heat of the 1930s, and discounting other warm, carbon-rich eras noted in the geological record, the cli-

---

25   Ibid.

mate change activists and influencers whip out their dramatized graphs to prove we are now just tenths of a degree from reaching the doomsday mark.

The solution?

Net zero, an over-the-rainbow scheme to replace fossil fuels with an electric grid 100 percent powered by renewable energy. It requires virtually every home and building in America to comply with sustainable standards including proper insulation and weathertight windows; replacing all gas appliances with electric; employing rooftop solar panels; and connecting all electrical features to a network that seamlessly interfaces with a smart meter. The corresponding result will be high energy prices, time-of-day power reductions, rolling blackouts, and the ability for an authoritative government to better control the behavior of the "lesser minds." But for the Davos elite and WEF's faithful stakeholders, nothing but opportunity lies ahead.

CHAPTER EIGHT

/////////////////////

ALTERNATIVE FANTASIES

"SOMEWHERE OVER THE RAINBOW..."

That song, penned by Yip Harburg and originally and beautifully recorded by Judy Garland in October 1938 for the 1939 film classic, *The Wizard of Oz*, allowed Garland's Dorothy to vocalize her deep yearning for a perfect, idyllic place somewhere outside of this world.

Today, the same song could serve as the theme for the pipe dreams and promises made to the climate cult, as those dreams and promises are pure fantasy about a utopia that doesn't really exist outside of their imaginations. With unrealistic ideas, such as powering the US electrical grid with nothing but renewable energy sources, it doesn't take long for someone looking through an intellectually honest lens to realize that the proposed ideas are based on slick rhetoric.

For example: on May 8, 2022, a beautiful Sunday afternoon, the state of California claimed to have set a historic milestone in its quest for clean, green, renewable energy. Up and down the Golden State's 840 miles of coastline and inland towards the fourteen-thousand-foot peaks of the Sierra Nevada, the sun was shining, and the winds were blowing moderately. Finally, a long cherished green goal was met: the state (supposedly) produced enough renewable electricity to meet 103 percent of consumer demand, which broke the record of 99.9 percent set just a week earlier.

This "accomplishment" made headlines around the world—understandably, since California not only possesses the United States' largest economy but also the fifth largest in the entire world.

However, the story was built on chicanery. The public was led to believe that the energy needs for California's forty million residents were seamlessly supplied by solar, wind, and renewables, and the holy grail—net zero—finally had been achieved.

But some key facts were intentionally left out of the report. More than 90 percent of all water heaters in California run exclusively on natural gas (as do 70 percent of all furnaces). Those features, though, are not considered a *direct* component of the electric energy grid. And the solar and wind output received critical assistance from their renewable cousins, hydropower and geothermal, not to mention another big boost from nuclear energy, none of which found their way into any of the media reports.

Once the sun drew closer to the horizon, and the natural diurnal winds ceased on that "record" day, the grid was supplied with natural gas to keep things running overnight and into the next day. In fact, it's common knowledge that all energy grids utilizing solar and wind need a reliable baseload backup plan, one that is running 24/7 just in case clouds hide the sun, and the wind is not blowing at optimum speed. Unless nuclear power is abundantly available, plan B has to be a fossil fuel, more often than not, natural gas.

Another important item the media missed was that May 8 was not only a Sunday (the day of the week with the least energy demand), but it was a day with comfortable temperatures averaging well below norms, thus, requiring much less need for air-conditioning. Add to the deception the fact that during the morning hours, and all day long in the mountains, gas furnaces or wood burning stoves kept otherwise chilly houses cozy.

California's energy ambition, codified in law, is to generate power completely free of greenhouse emissions by 2045. They also have a plan to never sell another gas-powered vehicle after 2035 and to ban natural gas furnaces and water heaters in all new construction by 2030. California already has some of the highest residential energy prices in the nation, and in the coming years, it will only get worse.

Thankfully, though, not everyone's brains have fallen out of their heads in Sacramento. An official state analysis of the state's aim to be 100 percent renewable makes it clear that natural gas must remain on tap.[1] The problem is that in their effort to be green, they have given the boot to nearly all their former in-state natural gas suppliers to the extent that now 90 percent of the natural gas used in California is imported from other states at high cost.[2]

On Monday, September 6, four months after their highly acclaimed climate triumph, California's temperatures soared into triple digits throughout much of the state, creating an energy demand so high that rolling blackouts were occurring to prevent the grid from crashing. Digital road signs above the freeways cau-

---

1   Liz Gill, Aleecia Gutierrez, Terra Weeks, California Energy Commission, "2021 SB 100 Joint Agency Report, Achieving 100 Percent Clean Electricity in California: An Initial Assessment," March 15, 2021, https://www.energy.ca.gov/publications/2021/2021-sb-100-joint-agency-report-achieving-100-percent-clean-electricity.

2   Laura J. Nelson, "The Gas Bill Is $907.13? Sticker Shock for Californians as Prices Soar," *Los Angeles Times*, February 15, 2023, https://www.latimes.com/california/story/2023-02-15/california-natural-gas-bills-expensive-socalgas-pge-long-beach.

tioned owners of electric vehicles: Do Not Charge EVs between 3 PM–9 PM.

The heat wave lasted three days (typical in California). Solar and wind alone could never come remotely close to keeping up with those needs during that period. Well over half the electrical energy used was delivered by natural gas.[3]

Everyone should pay heed to what happens in the Golden State on many environmental issues, because California has been the epicenter of the green movement from its inception, and its climate doctrine eventually ends up rolling across the country like a green tsunami.

CORPORATE WELFARE

We are all familiar with "the Midas touch," where everything touched turns to gold. The opposite of that could be said to be "the government touch," where almost everything the government inserts itself turns to rust. This is disastrously true in green investment.

Case in point, on May 26, 2010, as a large oil leak poured into the Gulf of Mexico, President Obama read from prepared notes regarding drilling:

> Part of what's happening in the Gulf is companies are drilling a mile underwater before they hit ground, and then a mile below that before they hit oil. With the increased risks, increased

---

3   Thomas Catenacci, "California's Grid Leaning Heavily on Natural Gas to Survive Energy Crisis, Despite Green Push," Fox Business, September 7, 2022, https://www.foxbusiness.com/politics/california-grid-leaning-heavily-natural-gas-survive-energy-crisis-despite-green-push.

costs—it gives you a sense where we're going. We're not going to be able to sustain this kind of fossil fuel use. This planet can't sustain it.[4]

The president's comments were not spoken from the shores of America's Gulf Coast but instead issued from the site of one of the biggest taxpayer-funded financial debacles (read *scams*) ever, the headquarters of Solyndra, a solar manufacturing firm located in Silicon Valley. The company had received a guaranteed government loan of over a half-billion dollars with terms like a quarterly interest rate of a ridiculously low 1.025 percent.[5] The deal also included a $25 million tax break from the state of California.

"It's here that companies like Solyndra are leading the way toward a brighter and more prosperous future," Obama proclaimed.

As the president went on to highlight the company's "cutting edge solar panels," insiders knew Solyndra's proprietary technology was *not* cutting edge but "bleeding edge." In other words, the product was so unique, there was a substantial risk of it being both unreliable and too expensive to use. Indeed, that proved to be one of the reasons the company failed.

Solyndra was Silicon Valley's most well-capitalized start-up company ever, having raised more initial funding than high-tech FAANG companies—Facebook (now Meta), Apple, Amazon, Netflix, and Google (now Alphabet). Conversely, Solyndra was

4   The White House, Office of the Press Secretary, "Remarks by the President on the Economy," Solyndra, Inc., Fremont, California, May 26, 2010, https://obamawhitehouse.archives.gov/the-press-office/remarks-president-economy-0.

5   ABC News, "'Connected' Energy Firm Got Lowest Interest Rate on Government Loan," September 7, 2011, http://abcnews.go.com/Blotter/solyndra-lowest-interest-rate/story?id=14460246.

also spending more money than any start-up ever. In its first five years, the firm burned through a billion dollars of private equity, using a substantial portion of it to build an exquisite 183,000-square-foot facility staffed by nearly one thousand employees. Once the government cash was tapped, the company rushed to develop an additional 609,000 square feet of manufacturing space, including shipping docks, office suites, an expansive staff cafeteria, and a commercial-size employee gym.

"When it's completed in a few months," Obama declared during his 2010 visit, "Solyndra expects to hire a thousand [additional] workers to manufacture solar panels and sell them across America and around the world."[6]

Shortly thereafter, the company received $10.3 million in long-term credit from the US Export-Import Bank to assist Solyndra's exports to Belgium.

A year later, the darling of sustainable development filed for bankruptcy, laid off its employees, and shuttered operations. In the end, American taxpayers footed the bill for Solyndra's shameful default on their $535 million loan guarantee from the Department of Energy (DOE).

Of course, early investors came out to be the big winners.

Solyndra's massive plant was purchased by Elon Musk's cousins for use by their company, SolarCity, chaired by Musk. It was a bottom-line business move that worked for a while. Not only did SolarCity get a great deal on the upscale plant, but they were also purchasing panels on the cheap from production overruns in China, installing them atop homes for free with residents leasing

---

6   The White House, Office of the Press Secretary, "Remarks by the President on the Economy," Solyndra, Inc., Fremont, California, May 26, 2010, https://obamawhitehouse.archives.gov/the-press-office/remarks-president-economy-0.

the panels while allowing SolarCity to sell excess energy back to utilities.

Eventually, though, the tide turned for SolarCity. Crippled with debt and struggling to turn a profit, Musk's cousins sold the company to his Tesla enterprise for $2.6 billion, considered a low valuation at the time. Now known as Tesla Solar, they manufacture electric vehicle chargers, roof solar infrastructure, and Tesla Powerwall batteries, the perfect fit for Tesla as they ride the alternative energy wave generated by climate change influencers and government mandates that make for corrupt corporate cronyism.

In my book, *Eco-Tyranny*, I detail quite a few environmental projects that went bust during the Obama years, but to summarize, when added up, the Obama administration oversaw about $100 billion in giveaways to wind and solar producers, electric car companies, and home and building weatherization assistance. It amounted to a huge corporate welfare experiment, enriching both industry and investors.[7]

The irony is, despite these awful losses, the Democrats do not give up. According to Joe Biden, "[Climate change is] the number one issue facing humanity. And it's the number one issue for me. Climate change is the existential threat to humanity. No one is going to build another oil or gas-fired electric plant. They're going to build one that is fired by renewable energy."[8]

The goal of the Democrat Party mimics that of California and the World Economic Forum in its embrace of net zero—a

7   Stephen Moore, "Obama's Green New Deal Was a Billion-dollar Bust," *Boston Herald*, March 27, 2019, https://www.bostonherald.com/2019/03/27/obamas-green-new-deal-was-a-billion-dollar-bust/.

8   Emma Newbuger, "Joe Biden Calls Climate Change the 'Number One Issue Facing Humanity,'" CNBC, October 24, 2020, https://www.cnbc.com/2020/10/24/joe-biden-climate-change-is-number-one-issue-facing-humanity.html.

meticulous scheme to "decarbonize" the energy grid. This is why Congress passed a $1 trillion energy infrastructure package signed into law November 2021. A big part of that plan is to have solar energy provide 40 percent of America's electricity by 2035, an astoundingly unrealistic figure given that since the days of Obama's presidency, solar power has only expanded from less than 1 percent of the nation's electricity portfolio to 3.4 percent today. Nonetheless, according to the Secretary of Energy, "This is truly a remarkable time for manufacturing in America, as President Biden's agenda and historic investments supercharge the private sector to ensure our clean energy future is American-made."

Well, of course it is a remarkable time, especially if you are in the renewables business.

LIKE ALCHEMY

From a scientific standpoint, solar energy *is* an amazing technology—about as close to alchemy as one can get—but its energy output is extremely expensive to produce. The most popular form of solar energy is known as photovoltaic, discovered in 1839 by French physicist Edmond Becquerel, who discovered that certain materials could produce small amounts of electric current when exposed to light—the "photovoltaic effect." In 1905, Albert Einstein further developed the concept upon which photovoltaic technology is based and received a Nobel Prize in physics for his efforts.

The first photovoltaic module was built by Bell Laboratories in 1954 and was considered a novelty and far too expensive to ever gain widespread use. In the 1960s, the US space program began to make the first serious use of solar technology to provide power aboard their spacecraft. Through the space program, photovoltaic solar advanced, and its reliability was established, though costs remained astronomical (pun intended). In 1979, as a global oil shortage occurred in the wake of the Iranian Islamic Revolution, environmental activists pounced on the crisis as a platform to demonize the entire oil and gas industry and encourage the use of solar power for residential applications.

The problem was, and still is, photovoltaic solar is an expensive proposition. Besides consisting of materials that must be mined (monocrystalline silicon, polycrystalline silicon, amorphous silicon, cadmium telluride, copper indium gallium selenide), there is the complex design of the high-tech panels, steep assembly costs, and pricey installation. Since the sun travels across the sky at different trajectories throughout the year, panels must be positioned at the best possible angles to catch the optimum direct radiation. If automated tilting panels are used to increase efficiency, then equipment costs are obviously higher. And ongoing maintenance must be factored in. A thin layer of dust or bird droppings will dramatically decrease the panels' energy-producing capacity, and degradation from harsh sun rays will eventually necessitate replacing the panels.

Plus, solar energy is pricey by the watt. A watt is a measure of electricity used by any power-consuming device. A thousand watts are known as a kilowatt. A kilowatt-hour (kWh) is the amount of energy consumed over the course of an hour. Utility companies use kilowatt-hours as the basis for what they charge their customers.

For example, a space heater rated at one thousand watts operating for one hour uses one kilowatt-hour of electricity. If a sixty-inch flat screen uses one hundred watts when it's turned on, then over the course of ten hours, it will consume one kilowatt-hour of power. An average electric clothes dryer consumes a kilowatt-hour of electricity every fifteen minutes. Add it all up, and the average household in the US consumes 886 kWh per month.[9]

Across the contiguous US, the per/kWh prices range from thirty cents in California to eleven cents in Louisiana, with the national retail average around fifteen cents per/kWh. So, on average, running that dryer for sixty minutes will cost someone about sixty cents.[10]

To arrive at the price of a kilowatt-hour of energy, your utility company looks at the various energy sources in their portfolio and carefully determines the cost involved to bring each product to market. This is called the "levelized cost." Calculations include everything from the price of raw materials (oil, natural gas, nuclear fuel rods), the construction costs of the power plants (including solar panels and wind turbines), ongoing maintenance, personnel, administration, generation, distribution, and so on.

After factoring all costs—not including government subsidies and tax breaks—the least expensive energy per/kWh is natural gas followed by coal, nuclear, wind, geothermal, and various solar applications, of which residential rooftop solar is the most expensive.[11]

---

9   US Energy Information Administration, Frequently Asked Questions, "How Much Electricity Does an American Home Use?," https://www.eia.gov/tools/faqs/faq.php?id=97&t=3.

10 US Energy Information Administration, Electric Power Monthly, "Table 5.6.A. Average Price of Electricity to Ultimate Customers by End-Use Sector," https://www.eia.gov/electricity/monthly/epm_table_grapher.php?t=epmt_5_6_a.

11 "Levelized Cost of Energy Comparison—Unsubsidized Analysis," Lazard, 2021, https://www.lazard.com/media/sptlfats/lazards-levelized-cost-of-energy-version-150-vf.pdf.

The biggest energy production problems for solar energy are clouds, dawn, dusk, and night hours. Without direct sunlight, the panels are in pause. This is why solar requires a full-time backup continually running at low levels in order to quickly ramp up as atmospheric conditions change. Again, most backup systems rely on fossil fuels.

It's a similar problem with wind turbines—they only generate electricity when the wind is blowing within certain speed parameters, and they, too, must have a full-time backup ready to take over at a moment's notice. A typical solar plant in the United States operates at only about 15 percent full capacity, and a wind plant at only about 25 percent, while gas and coal plants operate at 90 percent full capacity year-round. Thus, it takes about seven large solar plants and four massive wind farms to produce the same amount of electricity as a single coal-fired plant.[12]

This is why early on, the federal Energy Information Administration (EIA) issued a very frank warning about wind and solar technology:

> Wind and solar are intermittent technologies that can be used only when resources are available. Once built, the cost of operating wind or solar technologies when the resource is available is generally much less than the cost of operating conventional renewable generation. However, high construction costs can make the total cost to build and operate renewable generators higher

---

12　Charles Frank, "Why the Best Path to a Low-Carbon Future Is Not Wind or Solar Power," Brookings Institution, May 20, 2014, https://www.brookings.edu/articles/why-the-best-path-to-a-low-carbon-future-is-not-wind-or-solar-power/.

than those for conventional power plants. The intermittence of wind and solar can further hinder the economic competitiveness of those resources, as they are not operator-controlled and are not necessarily available when they would be of greatest value to the system.[13]

Solar energy is hardly a savior, and wind energy has its sizeable drawbacks too, especially given its real estate demands.

BLOWING IN THE WIND

It helps our understanding of the issue when we can discern the United States' current energy portfolio. The lion's share of energy usage is found in the transportation sector, which is powered by gas and diesel with a sliver coming from corn-based ethanol, a federal government mandate. In fact, about 36 percent of all the energy consumed in the United States is used by cars, trucks, planes, and trains.[14] There currently aren't enough electric vehicles on the road yet to effectively show up in the transportation energy analysis.

If we were to exclude transportation and focus only on types of energy used in the other sectors—industrial, residential, and commercial—the percentages[15] look like this:

13   US Energy Information Administration, *International Energy Outlook, 2010* (July 2010), 81, https://rosap.ntl.bts.gov/view/dot/5916.

14   "What Types of Energy Does the US Produce and Consume? How Much Energy Do Americans Use?" USA FACTS, 2022, https://usafacts.org/state-of-the-union/energy/.

15   US Energy Information Administration Information, "Electricity Explained," 2023, https://www.eia.gov/energyexplained/electricity/electricity-in-the-us.php.

Natural gas	39.8 percent
Coal	19.5 percent
Nuclear	18.2 percent
Wind	10.2 percent
Hydro	6.3 percent
Solar	3.4 percent
Biomass	1.3 percent
Petroleum	0.9 percent
Geothermal	0.4 percent

As you can see, renewable energy (wind, hydro, solar, bio, and geo) currently comprises less than 22 percent of America's energy output.

Now envision the actual energy footprint—literally the amount of land these sources of power generation each take up.

To start, two-thirds of the entire footprint is devoted to the corn grown for federally mandated ethanol (used in gas and diesel). This is "biomass" energy that totals fifty-one million acres, an area slightly larger than the state of Missouri. Next, because this source requires dams to hold back large bodies of water, hydropower takes up nearly nine million acres. Though wind is a smaller component in the portfolio, it has a significant land footprint at 6.5 million acres, while solar is a half-million acres.

Meantime, oil and gas sit on 3.5 million acres.

How much land is required to make way for all the required wind and solar farms?

Princeton University's Net-Zero America project, the source of our information, is a huge fan of abandoning fossil fuels. By their calculations, in 2050, when net zero is forecast to be running at full strength, a region equaling the size of Nebraska, Kansas, Oklahoma, Arkansas, Illinois, Kentucky, and southern Indiana is envisioned to be one humongous 250-million-acre wind farm. And solar energy would occupy 17.5 million acres, about all of northern Indiana.

But before we rush to make our national motto, In Wind We Trust, here is a caveat from Princeton University that should be considered:

> In this highly electrified economy, wind and solar provide four times the electric power capacity of the 2020 U.S. grid. Electricity powers all vehicles, heats homes and powers many industrial processes. When demand peaks and the grid needs an extra boost, it will come from a mix of batteries, hydropower, and combustion turbines burning carbon-free synthetic fuels and hydrogen.

This is nothing but a recipe for widespread rolling blackouts, constant government warnings to conserve energy, and the government's power to curb our energy output any time they deem it necessary via the smart grid.

SMART GRID

I received a lot of criticism when I wrote about an interactive grid that could override one's residential energy preferences in my book *Climategate*, but my assumptions then have all been proven to be correct. What I did not know in 2010, though, was that net zero was the next climate change fraud that would be coming down the pike.

But this actually goes back to the Federal Energy Act of 2005, signed into law by President G. W. Bush. Under the heading "Smart Meter," section 1252 of the act states, "It is the policy of the United States that…demand response shall be encouraged."

Demand response is a technology that gives an electric utility service provider the ability to override your home's electrical system and take control of your heating, air conditioning, and appliance usage. Section 1252 goes on to say:

> Not later than 18 months after the date of enactment of this paragraph, each electric utility shall offer each of its customers classes…a time-based rate schedule under which the rate charged by the electric utility varies during different time periods….[16]

This law presents a perfect example of how the government employs programs designed to change the public's behavior. It gave utility companies authority to offer customers a device to monitor energy usage in real time so they could be charged more when using electricity during periods of peak demand.

---

16 Addendum to Energy Policy Act of 2005, H.R. 6–370, "Smart Metering," Section 1252.

The monitoring equipment, technically known as advanced metering instrumentation (AMI), is popularly called the smart meter. It was designed to replace the spinning mechanism that was commonly located on the exterior wall of a home and observed monthly by a utility employee armed with a clipboard and canister of mace in case of unfriendly dogs. Early research led me to believe that these new meters would eventually give the service provider the ability to throttle back electricity or even turn off select devices in people's homes.

Turns out I got that one right too.

In 2008, California began pushing the roll out of the smart meters. I used my popular San Francisco radio program to rally listeners to urge the Energy Commission to reconsider the idea. It soon became a statewide rebellion, and the commission temporarily retired their plan. But as one staff member in Governor Arnold Schwarzenegger's office told me, the new meters were going to eventually happen because "it's part of a much larger plan."

And it sure was.

Today, almost ninety-eight million homes (69 percent) in the US are attached to a smart meter, as are 66 percent of all commercial buildings,[17] corroborating my hunch about the invasive remote-controlling capabilities. AMI provides two-way communication, allowing utilities to send information (including instructions and commands) to your home. This information can include time-based pricing information, demand-response actions, or even remote service disconnects. A fascinating research paper from the National Institutes of Health website focuses on

---

17 US Energy Information Administration, Frequently Asked Questions, "How Many Smart Meters Are Installed in the United States, and Who Has Them?," November 8, 2022, https://www.eia.gov/tools/faqs/faq.php?id=108&t=3.

consumer privacy concerns associated with the smart meter and concludes, "…consumers' privacy can be breached at any point."[18]

As we get closer to the projected enactment of net zero, advances in the smart meter, combined with further federal regulations, will enable the government to have a convenient big brother in your home to make sure we are compliant green citizens.

CALIFORNIA DREAMING

There is a chance, however, that net zero may tank long before 2050. It is an opinion based on all the other green fiascoes I have witnessed in California. Bumbling bureaucrats, massive cost overruns, duplicitous politicians with an eye for reelection, and early investors looking for easy money are a sure recipe for ruin. Just look at California's foolhardy high-speed rail line…to nowhere.

The electric "bullet train" has been in the works since 1996, the year the California High-Speed Rail Authority (CHSRA) was established. After four years of study, the commission determined that a high-speed rail system would be environmentally, economically, and socially profitable. In 2005, an implementation plan was approved, estimating the entire project would take eight to eleven years to complete. In 2008, voters approved a proposition providing $10 billion to fund a 380-mile high-speed rail line connecting San Francisco and Los Angeles. It was to be completed by 2020 with a total cost estimate of $33 billion. Eventually, other connections would be constructed around the

---

18 Jonathan Kua et al., "Privacy Preservation in Smart Meters: Current Status, Challenges and Future Directions," School of Information Technology, Deakin University, Geelong, VIC 3220, Australia, April 3, 2023, https://www.ncbi.nlm. nih.gov/pmc/articles/PMC10098615/.

state. In 2012, federal taxpayers forked over $3.3 billion in stimulus funding to the bullet train project.

Ironically, after years of delays due to environment lawsuits, in 2015, the train's supporters won a federal exemption from state environmental regulations, and, finally, a groundbreaking ceremony was held *in Fresno*—over two hundred miles away from the originally proposed north-south rail corridor linking LA and San Francisco. Governor Jerry Brown opened his remarks with a proclamation that was every bit as strange as the starting point location for this massive public works project: "What is important is the connection that we are rooted in our forebears and we are committed and linked to our descendants. And the high-speed rail links us from the past to the future, from the south to Fresno and north; this is truly a California project bringing us together today."[19]

The California High-Speed Rail Authority's chairman, Dan Richard, then stepped up to the mic, predicting, "We now enter a period of sustained construction on the nation's first high-speed rail system—for the next five years in the Central Valley and for a decade after that across California. This is an investment that will forever improve the way Californians commute, travel, and live. And today is also a celebration of the renewed spirit that built California."[20]

Three years later, a damning audit by the state concluded that flawed decision-making and poor contract management left

---

19 "California High-Speed Rail Authority Hosts Groundbreaking Ceremony," Metropolitan Transportation Committee, January 27, 2015, https://mtc. ca.gov/news/california-high-speed-rail-authority-hosts-official-groundbreaking-ceremony. (Governor Brown's quote is verbatim.)

20 Ibid.

the rail system fifteen years away from completion, with costs expected to be $77 billion.

In 2019, Governor Gavin Newsom gave the green light to continue the 171-mile section of track near Fresno with a projected cost of $22.8 billion and a completion date of 2030.[21]

In February 2023, with great bravado, the California High-Speed Rail Authority claimed to have reached a milestone, when in actuality, it inadvertently proved the project to be a complete boondoggle:

> **Fresno, Calif.**—The California High-Speed Rail Authority celebrated a historic milestone today, announcing the creation of more than 10,000 construction jobs since the start of high-speed rail construction. Most of these jobs have gone to Central Valley residents and men and women from disadvantaged communities.
>
> In partnership with the local and state Building and Construction Trades Council and the Fresno Regional Workforce Development Board, the Authority is proud to work alongside such skilled laborers, including electricians, cement masons, steel workers, and others who are helping bring the nation's first high-speed rail system to life while contributing to the local economy across five Central Valley counties.

---

21   Ralph Vartabedian, "New Cost Estimate for High-Speed Rail Puts California Bullet Train $100 Billion in the Red," *Cal Matters*, March 7, 2023, https://calmatters.org/economy/2023/03/california-high-speed-rail/.

"Ten thousand jobs created is one of many milestones to come on this historic project, and the Federal Railroad Administration remains committed to strengthening state partnerships to advance even more progress and deliver the passenger rail benefits people want and deserve," said administrator Amit Bose. "Today is an opportunity to celebrate jobs and what these jobs are creating. High-speed rail will revolutionize travel in California and contribute to a greener future, resulting in less congestion on our roads and at our airports."[22]

Ten thousand jobs for a train to nowhere while non-profit stakeholders like the Fresno Regional Workforce Development Board come out shining and union leaders smiling. A May 2023 update from the CHSRA, without fanfare, reported that the cost of the 171-mile segment (serving Fresno, Bakersfield, and Chowchilla) had shot up to $35 billion, more expensive than the original proposed rail line between Los Angeles and San Francisco. But even worse, with the entire project now estimated at $128 billion, politicians must find a spare $100 billion to finish the bamboozlement.

And furthering the lunacy, in July 2023 it was announced that a $20 million grant from the Federal Railroad Administration will help pay for restoration of downtown Fresno's historic Southern Pacific train depot, the money coming from the US Department of Transportation's Rebuilding American Infrastructure with

22   California High-Speed Rail Authority, "Photo Release: California High-Speed Rail Celebrates Creating 10,000 Construction Jobs," February 14, 2023, https://hsr.ca.gov/2023/02/14/photo-release-california-high-speed-rail-celebrates-creating-10000-construction-jobs/.

Sustainability and Equity (RAISE) program. According to the department's press release, "The project will also provide electric vehicle charging infrastructure and space for future transit charging in anticipation of the future California high-speed rail multimodal station."[23]

With such foolishness, waste, and fraud, it is difficult not to become cynical.

ELECTRIC VEHICLES

My personal favorite vehicle is the 2002 American-made four-wheel-drive pickup truck I drive. I've used it in all types of weather and terrain, and it's hauled lots and lots of stuff. The truck gets fewer miles to the gallon, but I don't mind; it's paid for itself many times over, and the important thing is, I enjoy it, as is my constitutional right to do so. Likewise, it's your right to own an electric vehicle. Many of the plug-in models are luxurious and super-fast. If you have one, I truly hope you enjoy your purchase. Just don't lecture me about how green it is.

In 2023, Tesla finally revealed its supply chain emissions[24] for the prior year: 30.7 million tons of carbon dioxide.[25] Given that the company sold 1.3 million vehicles in 2022, that is a significant emissions output. And, the fact is, electric cars are much more emis-

23 "High-Speed Rail Features in FY23 Raise Grants," Rail Passengers Association, June 28, 2023, https://www.railpassengers.org/happening-now/news/blog/high-speed-rail-features-in-fy23-raise-grants/.

24 Supply chain emissions include everything from the carbon footprint associated with business travel, to manufacturing the product, to getting the product to market.

25 Justine Calma, "Tesla's Carbon Footprint Is Finally Coming into Focus, and It's Bigger Than the Company Let on in the Past," *The Verge*, April 26, 2023, https://www.theverge.com/2023/4/26/23697746/tesla-climate-pollution-carbon-footprint-supply-chain-report.

sions-intensive to manufacture than gas or diesel vehicles. A big reason for the sharp contrast is because of the mining, refining, and manufacturing of raw materials needed to make the batteries for these vehicles. A Tesla's entire chassis, for example, is made up of over a thousand small batteries. Manufacturing your average sedan with an internal combustion engine creates about six metric tons of carbon dioxide emissions, while manufacturing an electric sedan creates more than ten metric tons of carbon dioxide emissions.[26]

Likewise, charging an electric vehicle generally requires the use of fossil fuels because, as stated, the primary source of energy in the grid is from fossil fuels; a fact that irritates many EV owners. As mentioned, the median household in America uses 886 kWh/month. Based on the US standard of fourteen thousand driving miles per year (1,166.6 miles per month), the average EV needs to be charged with 408 kWh/month.[27] That's almost half of an entire household's electricity demand for the same period. Less than 1 percent of all vehicles on the road are electric, but in 2022, 5 percent of all new vehicles sold were electric.[28] That translates to 762,883 brand new EVs or the equivalent of about 375,000 households.[29] Forget net zero. Our power grid is in desperate need of more traditional fuel sources to keep up with the

26 Liz Najman, "How 'Dirty' Are Electric Vehicles? The Answer Might Surprise You," *CBT News*, June 5, 2023, https://www.cbtnews.com/how-dirty-are-electric-vehicles-the-answer-might-surprise-you/.

27 Jacob Marsh, "How Many Watts Does an Electric Car Charger Use?" Energy Sage, August 4, 2023, https://news.energysage.com/how-many-watts-does-an-electric-car-charger-use/.

28 "How Many Electric Cars Are in the US?," Hertz, April 28. 2023, https://www.hertz.com/us/en/blog/electric-vehicles/how-many-electric-cars-are-in-the-us.

29 Zachary Shahan, "US Electric Care Sales Increased 65% In 2022," Clean Technica, February 2022, https://cleantechnica.com/2023/02/25/us-electric-car-sales-increased-65-in-2022/.

demand caused by our government's forced adaptation to electric cars and trucks.

A study from Stanford says, "Electric vehicles will contribute to emissions reductions in the United States, but their charging may challenge electricity grid operations."[30]

Most EV owners currently charge their vehicles at night, taking advantage of cheaper, off-peak electricity rates when demand is low, and fossil fuels (or nuclear power, depending on where you live) are providing most of the electricity. Right now, only 1 percent of all vehicles on the road are electric,[31] but if that number were to reach 30 percent, there would be major problems. The Stanford study indicates that in a net zero scenario, not only would the grid still require a fossil fuel back-up to keep from overloading, but it would also require massive, extremely expensive battery storage. Says the Stanford report:

> ...the needed grid storage requirements are substantial. Storage is expensive, current grid penetration is low, and the industry is already under pressure to scale up in the face of other grid challenges.[32]

Another environmental concern to consider, too, is that EV battery recycling is almost nonexistent. A friend who works at

30 Siobhan Powell et al., "Charging Infrastructure Access and Operation to Reduce the Grid Impacts of Deep Electric Vehicle Adoption," *Nature Energy*, September 22, 2022, https://www.nature.com/articles/s41560-022-01105-7.

31 Dustin Hawley, "What Percent of US Cars Are Electric?" JD Power, April 3, 2023, https://www.jdpower.com/cars/shopping-guides/what-percent-of-us-car-sales-are-electric.

32 Siobhan Powell et al., "Charging Infrastructure Access and Operation to Reduce the Grid Impacts of Deep Electric Vehicle Adoption," *Nature Energy*, September 22, 2022, https://www.nature.com/articles/s41560-022-01105-7.

the Tesla battery plant near Reno, Nevada, tells me there are countless used Tesla's being stored in hundreds of shipping containers in the desert surrounding the property. Perhaps because no one knows what to do with used-up batteries? And, when stored improperly, the batteries are a potential fire hazard. In fact, firefighters now undergo specific training for extinguishing EV fires, which require forty times more water than needed for a traditional car fire.

A research paper published by the Institute of Physics presents a grave warning:

> According to estimates, by 2025, 11 million tons of waste lithium-ion batteries will flood our market without a system to deal with them. If we are to deal with climate change, we must make full use of existing battery resources as much as possible to avoid pollution from toxic waste and ensure a strong supply of raw materials at low environmental costs. If discarded batteries cannot be effectively disposed of, it will cause huge damage to the environment and humans. When the battery is damaged, it will generate a lot of heat and cause a fire, and it will release incredibly toxic gas. In addition to humans, waste batteries have many potential hazards, and high concentrations of lithium can cause great harm to the human nervous system and endocrine system.[33]

33 Taotianchen Wan and Yikai Wang, "The Hazards of Electric Car Batteries and Their Recycling," IOP Conference Series: Earth Environment Science 1011 012026, 2022.

A study from Volvo reveals the environmental benefit of electric cars may never be felt because their production creates up to 70 percent more carbon emissions than their gas-powered equivalents. The plug-in vehicles must be driven tens of thousands of miles before they even begin to offset their higher manufacturing releases. If all the electrical charging needed came from solar and wind sources, Volvo says their C40 model still would have to be driven thirty thousand miles to break even, and with a traditional blend of electricity, it's actually closer to seventy thousand miles.[34]

However, many such vehicles will never hit those mileage targets, as owners upgrade to newer models, leaving loads of used electric cars sitting around unwanted. This is especially true for those who lease EVs. Customers typically turn the vehicle in at the end of the lease period and opt for a newer model.

Even if someone produces a viable plan to recycle batteries, millions of EVs will only stretch the grid, likely forcing utility companies to regulate when households with EVs will, or will not, be able to charge their miracle green machines.

Then there is the EV's sky-high price tag. The Argonne National Laboratory estimated that an average EV is about $22,000 more expensive to purchase than a comparable combustion vehicle. Sure, they cost about $14,000 less to fuel, insure, and maintain over a fifteen-year period, but their lifetime cost is still $8,047 more than a traditional vehicle.[35] However, as revealed

---

34 Sam D. Smith, "Volvo Says Manufacturing an Electric Car Generates 70 Percent More Emissions Than Its ICE Equivalent," *Car Scoops*, November 12, 2021, https://www.carscoops.com/2021/11/volvo-says-manufacturing-an-electric-car-generates-70-percent-more-emissions-than-its-petrol-equivalent/.

35 Brent Bennett and Jason Isaac, "Overcharged Expectations: Unmasking the True Costs of Electric Vehicles," Texas Policy Foundation, October 2023, https://www.texaspolicy.com/overcharged-expectations-unmasking-the-true-costs-of-electric-vehicles/.

in an associated paper published by the Texas Policy Foundation (based on 2021 data), without the massive government subsidies that are being doled out to encourage electric vehicle adoption, your average EV would cost $48,698 more to own over a ten-year period. That's because of the $22 billion in government favors given to EV manufacturers and owners.[36] Additionally, when adding the costs of the subsidies to the true cost of fueling a plug-in, it equates to an EV owner paying $17.33 per gallon of gasoline. And these estimates don't include the billions and billions of subsidies the Inflation Reduction Act offers to the EV supply chain, particularly for battery manufacturing.

"Somewhere over the rainbow," indeed.

CONCLUSION

The net zero research from Princeton is quite straightforward. If the US is to have a carbon-free economy by 2050 but does not want to invest in the land equivalent of seven states to do it, then we will "need to rely far less on wind and solar and instead build hundreds of nuclear plants and natural gas plants outfitted with systems to capture the carbon dioxide before it escapes into the atmosphere."[37]

With that alternative plan, wind and solar would contribute 44 percent of generated electricity, emission-free nuclear and natural gas power plants with pricey carbon-capture technology would provide another 50 percent, with the remaining 6 percent coming from hydro and geothermal sources. Under that situa-

---

36 Ibid.
37 Eric Larson et al., "Net-Zero America: Potential Pathways, Infrastructure, and Impacts," Princeton University's Net-Zero America Program, October 29, 2021.

tion, the US would need to build 250 large nuclear plants[38] and three hundred new natural-gas-fired power plants with a network of carbon-capture pipelines and carbon storage facilities. Natural gas and nuclear energy are very compact power sources. A conventional one-gigawatt reactor operating on one thousand acres produces the same amount of energy as a wind farm spanning one hundred thousand acres.

Needless to say, expanding nuclear power is a nonstarter for environmental activists.

On the other hand, if the US were to attempt to build a grid with 80 percent of its energy coming from solar and wind, an unfathomably huge battery backup storage system would be required. According to the genius minds at MIT, twelve hours of battery storage for such a grid would cost more than $2.5 trillion dollars.[39]

And California's now-infamous train to nowhere perfectly illustrates that point so poignantly: money somehow never hinders the climate agenda's high priests from adamantly continuing to push one of the most underhanded schemes in the history of mankind.

---

38 Each reactor would require a capacity of at least one gigawatt, or they recommend several thousand smaller modular reactors.

39 James Temple, "The $2.5 Trillion Reason We Can't Rely on Batteries to Clean Up the Grid," *MIT Technology Review*, July 27, 2018, https://www.technologyreview.com/2018/07/27/141282/the-25-trillion-reason-we-cant-rely-on-batteries-to-clean-up-the-grid/.

CHAPTER NINE

WEATHER GODS

1.5° CELSIUS—THE ATMOSPHERIC WARMING THRESH OLD. The tipping point. The mercury spike that will shatter the earth's thermometer forever and bring a sweltering end to all life on earth. That dreaded 1.5° rise in temperature must be averted—at all costs.

Or so go the invocations of the maniacal high priests of the climate cult.

Their climate change catechism contends that since the last Ice Age, some ten or twelve thousand years ago, the earth was naturally moderating until humans discovered evil fossil fuels, an abundant source of cheap energy that enabled capitalists to exploit labor and launch a revolution—the Industrial Revolution. And after the births and deaths of billions of humans, along with their countless polluting cars, trucks, buses, trains, planes, large houses, office buildings, stores, paved streets, and parking lots for their personal vehicles—the earth's temperature has now risen by 1.2° Celsius. Any day it could reach the doomsday mark. That is why the world's citizens must adopt and conform to net zero. And if that does not reverse the cursed thermometer, more drastic measures must be taken. They must. All the computer models insist on it.

BRIAN SUSSMAN

And more drastic measures have already been in the works for quite a while.

If the sound scientific evidence (much of which you have now seen for yourself) didn't reveal otherwise, this could all sound like an outrageous CGI sci-fi movie trailer. But the reality is, it's actually a slick propaganda pitch designed to coerce us into government compliance, and in far too many cases, it is working amazingly well. Our schools have continued to churn out multitudes of students who have been instructed *what* to think, not *how* to think, for at least three generations.

Ask someone who has been indoctrinated by the climate change propaganda about the 1.5° disaster zone we are allegedly fast approaching. "With the world temperature already warmed up by one point two degrees Celsius toward disaster, will we even notice another zero point three–degree increase?"

Most likely his or her response will be a rote, Google-like explanation that "even a small rise in temperature will increase the frequency and intensity of extreme weather events around the world, including droughts, storms, wildfires, and heatwaves."

"But what about extreme cold temperatures?"

"The earth is currently experiencing the warmest weather on record," they reply, avoiding the extreme cold question altogether.

And if you press them further on the issue, the response might be a curt, "There is no extreme cold."

Even though, just last year, Mount Washington, New Hampshire, reported a windchill of minus 108° Fahrenheit, the coldest temperature *ever* recorded in the United States.[1] And

---

1 February 17, 2023.

222

California recorded its fifth coldest March since 1885.[2] And that's just the United States.

In 2021, the United States Antarctic Program of the National Science Foundation and the National Snow and Ice Data Center reported the average temperature at the South Pole between their winter months of April and September was minus 78° Fahrenheit, the coldest on record. So cold, the extremes pushed sea ice levels surrounding the continent to one of their highest-ever levels.

He or she isn't buying it, so, you give it one last shot. "A Nobel Prize–winning physicist says the so-called 'climate emergency' claims are a 'dangerous corruption of science that threatens the world's economy and the well-being of billions of people.'"[3]

You're met with silence.

The cult is not interested in facts.

By now, it should seem apparent that climate change adherents of all stripes—from innocent grade-schoolers to the PhDs— usually have no time for debate. And why should they? They remain smugly confident in their dogma or prophetic belief that the world is tumbling wildly toward the 1.5° extinction threshold. The ice will melt, causing the oceans to rise and submerge the islands, that will generate hurricanes that will fill the atmosphere with rivers of water that will drop rain bombs from the sky, causing world-wide flooding and upheaval.

Yet even when an eminent scientist like Dr. John Clauser, winner of *two* Nobel Prizes in physics, steps forward to proclaim

2   March 2023 was the fifth coldest March in California since 1895. The average temperature across the state was 44.2 degrees Fahrenheit, nearly 9 degrees lower than the March average of 53.1 degrees.

3   "Nobel Laureate John Clauser Elected to CO_2 Coalition Board of Directors," CO_2 Coalition, May 5, 2023, https://co2coalition.org/publications/nobel-laureate-john-clauser-elected-to-co2-coalition-board-of-directors/.

that the climate agenda is being driven by "massive shock-jour-nalistic pseudoscience,"[4] they plug their ears and refuse to listen.

Tone deaf or truth deaf, the climate change leaders con-tinue plowing forward with plans frighteningly beyond net zero, despite the potentially dangerous consequences.

CLIMATE MODELS: GARBAGE IN, GARBAGE OUT

Dr. Clauser's resume is stellar. He earned a bachelor of science degree in physics from Caltech and a master's and doctorate in physics from Columbia University. From 1969 to 1996, he worked at Lawrence Berkeley National Laboratory, Lawrence Livermore National Laboratory, and the University of California, Berkeley. And as mentioned, he has received two Nobel Prizes in his distinguished field.

An authentic scientist's scientist, he contends there is no real climate crisis:

> The popular narrative about climate change reflects a dangerous corruption of science that threatens the world's economy and the well-being of billions of people. Misguided climate science has metastasized into massive shock-journalistic pseudoscience. In turn, the pseudoscience has become a scapegoat for a wide variety of other unrelated ills. It has been promoted and extended by similarly misguided business marketing agents, politicians, journalists, government agencies, and environmentalists. In my opinion, there is no real climate crisis.[5]

4   Ibid.
5   Ibid.

Dr. Clauser developed a climate model that provides a vital dynamic that the current simulations—the ones used to forecast the 1.5° powder keg—have totally ignored.

The atmospheric layer just above the earth, where weather occurs, is the troposphere. It averages about forty thousand feet high—almost eight miles—and gets its name from the Greek word, *tropos*, meaning "turn," so named because the lower portion of the troposphere is where air is warmed by the earth's surface and rises, promoting vertical mixing or turning. The troposphere contains roughly 80 percent of the mass of the earth's atmosphere.

The layer above troposphere is the stratosphere. It extends from the top of the troposphere to an altitude of about 180,000 feet—thirty-four miles up. In contrast to the troposphere, which experiences decreasing temperatures with altitude, stratosphere temperatures increase closer to the sun. This vital layer possesses ozone, which absorbs ultraviolet radiation from the sun, protecting life on the earth from too much heat. There is little mixing or turning in the stratosphere and scant weather, though clouds do form near the poles and giant thunderheads regularly burst upward into the region.

The towering tropospheric cumulus clouds and thunderstorms (which produce one hundred lightning flashes on earth every second) and all the other various clouds scattered across the skies play a preeminent role in the earth's colossal climate system. The clouds, constantly churning and morphing, reflect the sun's energy back into space and trap heat rising from the surface. It's an awesome process that I contend is divine. And therein lies a significant problem for the climate change cultists: the climate models cannot accurately account for clouds, thunderstorms, or even water vapor, which is the most abundant greenhouse gas in our atmosphere.

There's also another inherent shortcoming in the models in that they are unable to accurately replicate the powerful ocean currents that act as massive weather conveyor belts, moving massive air masses and storms all around the earth. These currents naturally regulate the earth's temperatures by transporting the waters warmed by the sun in the tropics to the poles and back again.

The ocean's currents are incredibly complex, driven by a combination of variable water temperatures, salinity gradients, surface winds, the earth's rotation, and the gravitational effects of the sun and moon, causing tides. Adding to this complexity is the fact that the currents typically flow clockwise in the Northern Hemisphere, counterclockwise in the Southern Hemisphere, and change course when they cross the equator. This constant, ongoing circulation assists in counteracting the uneven distribution of solar radiation that reaches the earth's surface.

But all these fascinating fine points are impossible to properly input into the mathematical computer simulations. Hence, the need to approximate. And what cannot be approximated is simply omitted.

One must at least wonder if personal bias or political agendas ever come into play when a scientist approximates or omits information in the model. It's akin to solving all the world's problems with a pen on the back of a cocktail napkin.

The fact is, climate models are known to greatly underestimate the feedback from clouds and, thus, have a poor record of accomplishment. Apart from failing to predict the lack of global warming in the early 2000s, the models are known to consistently run hot.

As a former weather forecaster, I can say with certainty that a prediction for the next day is about 90 percent correct when it comes to a *general* description of what's coming (sunny, hot, cold, windy, rain, snow, and so forth). A five-day forecast is broadly accurate about 70 percent of the time if you are looking for the specific high and low temperature and an overall idea of what the weather will be like. But by the time you get to the ten-day forecast, things get giddy quickly. And yet we are allowing humanity's future to be determined by politicians and elite influencers who provide funding for all the imperfect scientists armed with expensive computers, and who, in turn, create climate models that spell existential gloom and doom from every direction. Adding further to the irony is the fact that the same imperfect models are now looking to other inexact simulations to try to produce a cure for the climate's supposed ills.

Astronaut Harrison Schmitt, the PhD geologist who walked on the moon, spoke about the models at a conference of professional science writers:

> I'm a geologist. I know the earth is not nearly as fragile as we tend to think it is. It has gone through climate change. It is going through climate change at the present time. The only question is, is there any evidence that human beings are causing that change? Right now, in my profession, there is no evidence. There are models. But models of very, very complex natural systems are often wrong.

The observations we make as geologists and observational climatologists do not show any evidence that human beings are causing this. Now, there is a whole bunch of unknowns. We don't know how much CO_2, for example, is being released by the southern oceans as the result of natural climate change that's been going on now since the last ice age…So, the only skepticism I have is, is what is the cause of climate change?

…I, as a scientist, expect to have people question orthodoxy. And we always used to do that. Now, unfortunately, funding by governments, particularly the United States government, is biasing science toward what the government wants to hear…That's a very dangerous thing that's happening in science today, and it's not just in climate. I see it in my own lunar research. If NASA's interested in a particular conclusion, then that's the way the proposals come in for funding. So, it's a very, very serious issue, and I hope the science writers in this room will start to dig deeply into whether or not science has been corrupted by the source of funds that are now driving what people are doing in research, and what their conclusions are.[6]

---

6   "Apollo Plus 50: Past and Future of America's Space Effort," Council for the Advancement of Science Writing, ScienceWriters Conference, 2018, YouTube, https://www.youtube.com/watch?v=heE7a6xs6QY.

A forthright article at MIT's Climate Portal provides a succinct closure to Dr. Schmitt's comments: "The earth's climate is too complex for even the most powerful computers to fully simulate."[7]

Dr. Clauser's new model properly accounts for the clouds and other inconsistencies, providing him the confidence to state unequivocally, "There is no real climate crisis."

RAINMAKERS

The question is, are those driving the climate agenda willing to consider Dr. Clauser's conclusions?

My experience says, "No."

As demonstrated, this agenda is being driven by a consortium of organizations and governments who seek to dismantle life, liberty, and the pursuit of happiness anywhere such ideals are found. And the insidious tangent of this agenda, the one that keeps me up at night, is known as geoengineering. It is a topic that too often is quickly dismissed as a conspiracy theory associated with contrails. But that is a false accusation. The climate agenda really has plans to create and manipulate weather to transform the global atmosphere.

They have posited themselves as the weather gods.

Weather *can* be modified, even on a large scale. The US discovered the procedure during World War II and has been refining it ever since, as I will explain. Currently, some fifty countries are engaged in weather and climate adaptation—China boasts employing some thirty-seven thousand people engaging in

---

7   Efatih Eltahir, "Climate Models," Climate Portal, Massachusetts Institute of Technology, January 28, 2021, https://climate.mit.edu/explainers/climate-models.

weather warfare on its behalf. Most commentary describes geo-engineering as a necessary tool to overcome the current effects of climate change, like artificially overcoming drought by manifesting rainfall for raising crops.

But the available literature also speaks of another type of geoengineering that many prominent voices contend is critical to keep us under 1.5°—solar geoengineering. Its most popular derivative, stratospheric aerosol injection (SAI), involves releasing particulates into the atmosphere to reduce the amount of sunlight hitting Earth's surface. The idea was inspired by clouds of ash produced by volcanic eruptions, specifically the Mount Pinatubo eruption in 1991 that contributed to a global cooldown over the following few years.

Geoengineering had its inception in the mind of Nobel Prize-winning American physicist, chemist, and engineer, Dr. Irving Langmuir, during World War II. Along with Dr. Vincent Schaefer, Langmuir was employed by General Electric (GE) and tasked with finding a way to de-ice military aircraft, create aerial smoke screens, and even generate weather modification. On November 13, 1946, Schaefer was flown by pilot Curtis Talbot to an altitude of fourteen thousand feet, about thirty miles east of Schenectady, New York, to conduct an experiment: make snow.

In a brief instructional film produced by GE a year later, Schaefer described the details of the experiment. Six pounds of dry ice (frozen carbon dioxide) were delivered into a four-mile-wide cloud formation. As they turned back to look at the clouds, Dr. Schaefer noted, "Within a very few minutes we saw long streamers of snow falling from the base of the cloud and evaporating through the drier air below." With much enthusiasm, he continued, "Under many conditions, of course, full-fledged snowstorms will be produced in this way. Nature, at last, has permitted us to do a little something about the weather."[8]

Schaefer's experiment verified a hypothesis, originally advanced in 1933 by Swedish meteorologist Tor Bergeron and German physicist Walter Findeisen, that clouds would precipitate if they contained the right mixture of ice crystals and supercooled water drops.

While common dry ice was the original weather-making ingredient, further experimentation discovered that silver iodide was even more effective for making weather. Soon the first indications of something far greater than just creating precipitation became apparent. The possibilities led the military to support a broad theoretical laboratory and field program in cloud modification from 1947 to 1952 known as Project Cirrus. Dr. Langmuir was involved in the project, which included 181 air missions. In Cirrus's official summary notes, he states, "Such work could easily have profound economic, political, and military effect."

Langmuir soon became recognized as the pioneer in "rainmaking." An article in the *Charleston Daily Mail* (Charleston, WV) referred to his discovery as comparable to an atomic weapon:

> "Rainmaking" or weather control can be as powerful a war weapon as the atom bomb, a Nobel prize winning physicist said today.
>
> Dr. Irving Langmuir, pioneer in "rainmaking," said the government should seize on the phenomenon of weather control as it did on atomic energy

---

8   Museum of Innovation and Science (miSci), "Dr. Vincent Schaefer Snow-Making Demonstration," YouTube, https://www.youtube.com/watch?v=2D5s2FlA_5k.

when Albert Einstein told the late President Roosevelt in 1939 of the potential power of an atom-splitting weapon. "In the amount of energy liberated, the effect of 30 milligrams of silver iodide under optimum conditions equals that of one atomic bomb," Langmuir said.[9]

In August 1953, the US formed the President's Advisory Committee on Weather Control. Its stated purpose was "to make a complete study and evaluation of public and private experiments in weather control for the purpose of determining the extent to which the United States should experiment with, engage in, or regulate activities designed to control weather conditions."[10]

In 1958, the committee's chairman, Captain Howard Orville, a graduate of the Naval Academy with an advanced degree in meteorology from MIT, warned, "If an unfriendly nation solves the problem of weather control and gets into the position to control the large-scale weather patterns before we can, the results could be even more disastrous than nuclear warfare."[11]

A year earlier, in 1957, Lyndon B. Johnson ("LBJ") was the Senate majority leader and leader of the Democratic political opposition to Republican president Dwight Eisenhower. LBJ forced Eisenhower into a space race he did not want after the

9   Matt Novak, "Weather Control as a Cold War Weapon," *Smithsonian Magazine*, December 5, 2011, https://www.smithsonianmag.com/history/weather-control-as-a-cold-war-weapon-1777409/.

10 "Final Report of the Advisory Committee on Weather Control," *Journal of the American Meteorological Society*, vol. 39, no. 11 (November 1958), https://journals.ametsoc.org/view/journals/bams/39/11/1520-0477-39_11_583.xml.

11 Matt Novak, "Weather Control as a Cold War Weapon," *Smithsonian Magazine*, December 5, 2011, https://www.smithsonianmag.com/history/weather-control-as-a-cold-war-weapon-1777409/.

USSR got the jump on the US with the launch of Sputnik 1, the first artificial space satellite.

LBJ had big technological ambitions. On May 27, 1962, as John F. Kennedy's vice president, he presented a stirring graduation address at his alma mater, Southwest Texas State University. Besides speaking of technology that would someday "permit the people of the world to watch one television program at the same time throughout the world," he then boldly declared that such innovation would lay the "predicate and the foundation for the development of a weather satellite that will permit man to determine the world's cloud layer, and ultimately to control the weather, and he who controls the weather will control the world."[12]

As president, in 1965, LBJ composed a Special Commission on Weather Modification that produced a 149-page report supporting the continuation of exercises to control the weather:

> It should be clear that a long-range program of weather and climate modification can have a direct bearing upon the main purposes of American foreign policy. It can contribute to defending the security of the United States and other nations of the free world. It can aid the economic and social advancement of the developing countries, many of which face problems associated with adverse climatic conditions and serious imbalances in soil and water resources....

---

12 "Vice President Johnson at Southwest Texas State University (1962)," *Texas Archive*, https://texasarchive.org/2010_00003.

The challenge and the opportunity presented to
the world community by the prospect of man's
achieving the ability to modify the atmospheric
environment form one of the most exciting long-
range aspects of the subject. It involves the possible acquisition of a new and enormous power
to influence the conditions of human life. The
potentialities for beneficial application are vast,
as are also the potential dangers.[13]

In October 1966, during the Vietnam War, the Johnson
administration authorized a military weather modification exercise that worked just as anticipated. "Project Popeye" launched
more than fifty cloud-seeding flights to disperse particulate above
the routes used to bring communist troops and supplies into
South Vietnam. Though the project could likely impact neighboring Laos, the Laotian government was never consulted nor
warned. A 1967 memorandum[14] addressed to US secretary of
state Dean Rusk illustrated the results:

(a) 82% of the clouds seeded produced rain within a brief
    period after seeding—a percentage appreciably higher
    than normal expectation in the absence of seeding.

13 National Science Foundation, "Weather and Climate Modification, Report of the
    Special Commission on Weather Modification," NSF-663, December 20, 1965,
    https://www.nsf.gov/nsb/publications/1965/nsb1265.pdf.

14 "Memorandum from the Deputy Under Secretary of State for Political Affairs
    (Kohler) to Secretary of State Rusk," Foreign Relations of the United States,
    1964–1968, vol. 28, Laos, 274, January 13, 1967, https://history.state.gov/
    historicaldocuments/frus1964-68v28/d274.

(b) The amount of rainfall induced by seeding is believed to have been sufficient to have contributed substantially to rendering vehicular routes in this area inoperable. Since the end of the rainy season, the communists have failed to undertake route repairs and there has been no vehicular traffic.

(c) In one instance, the rainfall continued as the cloud moved eastward across the Vietnam border and inundated a U.S. Special Forces camp with nine inches of rain in four hours.

True to LBJ's predictions, the US proved weather could be created and controlled. Other countries would soon enter the race to catch up with the world's superpower.

SKY WARS

In August 2008, shortly before the opening ceremonies of the Summer Olympics in Beijing, the sky was obscured by thick smog. Not a good backdrop for modern China's televised debut on the world stage, but the communist regime was ready. Their Weather Modification Office was prepared to spare the air above the city and its twenty million residents. Aircraft were deployed to seed the skies with silver iodide, while technicians manned modified antiaircraft guns to fire canisters of the same compound into the clouds. On cue, ice-like crystals formed, water vapor condensed, and it rained in and around the city, washing away the smog from the sky above.

China claims to have become quite adept at controlling the weather. Today, tens of thousands are employed in rainmaking projects across the nation. Cloud seeding is so normal in China that on smog free days in the larger cities, locals suppose an important political gathering must be taking place.

And be sure, this is not innocent tinkering. Geoengineering beyond simple dry ice or silver iodide is fraught with risk. Weather modification is occurring all around the world through aerosol dispersal of toxic metals and harmful chemicals, and that, combined with the worldwide grid of High-Frequency Active Auroral Research Program (HAARP) technology, poses a real and frightening threat to all humankind. If true, these bioweapons may be causing an increase in respiratory and skin diseases. There is much speculation regarding who may be behind this weather modification and why, but two facts are glaringly obvious: 1) chemical spraying from planes is happening everyday around the world, and 2) the governments of the world, including our own, are not being forthright. And these dangers don't even include the risks associated with weaponizing the technology for weather warfare to "control the world."

The stratosphere, too, can be manipulated. If seeded with particulate matter, aluminum oxide or sulfur dioxide (SO_2), artificial clouds can be formed in the stratosphere to reflect sunlight and radiation away from the earth to cool the planet. This is known as "solar geoengineering."

So, what if a country—or wealthy individual—unilaterally decided to act and deploy aerosols into the atmosphere to reduce the temperature? Who knows how such spraying could throw the climate completely out of whack, let alone the harmful health effects from disbursing such toxins?

A dramatic research paper, "Owning the Weather in 2025," was prepared for the Air Force in 1996.[15] The paper details topics such as storm enhancement (decreasing human comfort levels), precipitation denial (inducing droughts), precipitation avoidance (improve comfort level), and cloud removal. The research also speaks of the "alteration of global climate on a far-reaching and/or long-lasting scale," something the authors imply was not possible in 1996, but once available, would be considered "potential military options, despite their controversial and potentially malevolent nature and their inconsistency with standing UN agreements to which the US is a signatory."

Even the Biden administration's 2021 National Intelligence Estimate on climate change sounded the alarm about these risks, highlighting the "increasing chance that countries will unilaterally test and deploy large-scale solar geoengineering—creating a new area of geopolitical disputes."[16]

There are currently many players conducting theoretical research and too many voices on the sidelines encouraging action. While small corners of the internet contend dangerous geoengineering is presently occurring, complete with compelling videos and atmospheric test results suggesting dangerous fallout from the various aerosols prescribed for use in solar geoengineering actions, most of the buzz is quite enthusiastic about the potential for taking a major step forward to go "where no man has gone before," to turn a popular sci-fi phrase.

15 Col. Tamzy J. House et al., "Weather as a Force Multiplier: Owning the Weather in 2025," April 1996, https://apps.dtic.mil/sti/pdfs/ADA333462.pdf.
16 "Climate Change and International Responses Increasing Challenges to US National Security through 2040," *National Intelligence Estimate*, 2021.

Unsurprisingly, a compelling urge to do more than just net zero is growing. It has become impossible to watch, read, or listen to the news without being force-fed a steady diet of climate change disinformation. A guest blogger at the globalist Council on Foreign Relations website pleads:

> Drastic action is needed before the window of opportunity closes altogether. Given that emissions reductions alone are likely to prove insufficient in preventing severe effects from climate change, scholars and policymakers are looking to geoengineering techniques as a possible last resort. Injecting particulate matter into the lower stratosphere, for instance, could help cool the planet by reflecting sunlight into space...Failing to study and develop geoengineering methods to help prevent the severe effects of climate change would be irresponsible.[17]

The progressive Brookings Institution also makes an urgent appeal for solar geoengineering:

> As worsening climate conditions are likely to generate continued worldwide interest in geoengineering deployment, the world cannot afford to wait until an emergency arrives if it hopes to navigate increasingly complicated climate and geoengineering scenarios. States should collaborate to develop

---

17 Terrence Mullan, "The World May Need Geoengineering, and Geoengineering Needs Governance," *The Internationalist*, Council on Foreign Relations, July 25, 2019, https://www.cfr.org/blog/world-may-need-geoengineering-and-geoengineering-needs-governance.

and refine international norms and institutions to adequately prepare the world for the inevitable stresses of continued climate change and the resulting reaction from states and other actors.[18]

WEATHER GODS

Recall Marx's laws of matter. According to that false doctrine, there are some who have been born with greater intellect than most, and those from such stock have a responsibility to keep the lesser-minded in check to keep them from destroying both humanity and the planet.

Three billionaires who seem to believe they are from that elitist pedigree are George Soros, Bill Gates, and Jeff Bezos. All of them have made moves to save the planet via solar geoengineering.

In 2023, at the age of ninety-two, Soros took the stage at the Munich Security Conference to expound upon the risk of climate change. His solution: reflect the sun's energy away from the polar ice caps to keep them from melting.

> Let me start with a bold assertion…our civilization is in danger of collapsing because of the inexorable advance of climate change. This is a very succinct statement, but I believe it provides an accurate summary of the current state of affairs.
>
> [British climate scientist] Sir David King has a plan to repair the climate system. He wants to

18 Joseph Versen et al., "Preparing the United States for Security and Governance in a Geoengineering Future," *Brookings*, December 14, 2021.

recreate the albedo effect by creating white clouds high above the earth. With proper scientific safeguards and in consultation with local indigenous communities, this project could help re-stabilize the Arctic climate system which governs the entire global climate system.

The message is clear: human interference has destroyed a previously stable system and human ingenuity, both local and international, will be needed to restore it.

At present, practically all the efforts to fight climate change are focused on mitigation and adaptation. They are necessary but not sufficient.

The climate system is broken, and it needs to be repaired. That's the main message I'd like to convey this evening.[19]

Then there is Bill Gates. In 2021, he financially backed the Harvard University Solar Geoengineering Research Program. Harvard is a bastion for belief in this technology. The school's student body website explains what the future of solar geoengineering will look like:

...it would entail flying a suite of aircraft 30-odd miles above earth's surface to inject millions of tons of sulfate particles into the air. High up in the stratosphere, these particles would shroud the

---

19 George Soros, "Remarks Delivered at the 2023 Munich Security Conference," GeorgeSoros.com, February 26, 2023.

globe in a chemical mirror, reflecting away some
of the sun's radiation before it could be trapped
by greenhouse gases.[20]

Harvard's cheery summary reads like a chant from the bleach-
ers, "Go, team, go!"

After reading some of the work procured by Harvard's
Geoengineering Research Program, it seems like they are trying
to downplay their efforts, continually using the word "small" to
minimize perceptions about the atmospheric intervention they
intend to induce high in the sky. In one paper, they insist what
they are doing "is not a test of solar geoengineering per se."[21]

The program initially aimed to send up a research balloon
in Sweden, release a "small amount" of calcium carbonate and
sulfuric acid into the stratosphere, and then measure the amount
of light scattered away from Earth. However, shortly after going
public with the proposed experiments, protestors rose up, and
Sweden shut the plans down.

Jeff Bezos has also entered into the fray with his billions of
dollars, putting Amazon's supercomputers to work to model
the effects of SO_2 injections into the atmosphere; and Dustin
Moskovitz, a billionaire Facebook cofounder, pumped in nearly
$1 million to study the potential effects of solar geoengineering.

But the biggest funding pipeline seems to be coming from
the US taxpayers. In the Consolidated Appropriations Act,
passed by Congress in 2022, the White House Office of Science

---

20 Maliya V. Ellis and Saima S. Iqbal, "Is It Time to Consider Dimming the Sun?"
    The Harvard Crimson, October 28, 2021.

21 Keutsch Group at Harvard, "SCoPEx: Stratospheric Controlled Perturbation
    Experiment," https://www.keutschgroup.com/scopex.

and Technology Policy (OSTP), in coordination with relevant federal agencies, has been commissioned to develop a five-year "scientific assessment of solar and other rapid climate interventions in the context of near-term climate risks and hazards." The research plan will assess the various climate interventions, including spraying aerosols into the stratosphere, and what impact these kinds of climate interventions may have on Earth.

Spraying sulfur dioxide into the atmosphere is known to have harmful effects on both the environment and human health. Of course, geoengineers contend we must determine how to balance these risks against a potential catastrophic rise in the earth's temperature. One such scientist is Dr. David Keith, who has written a book, *The Case for Climate Engineering*. Keith is a professor of physics who formerly ran Harvard's geoengineering research and now does similar work at the University of Chicago. Keith hit the public's radar during a 2013 interview on Steven Colbert's late night television show, rationalizing the additional deaths caused by geoengineering in an effort to save the planet. His analogy was based on stopping global warming in seven years. Here is a partial transcript of that interview:

> **Keith:** You start with a fleet of just two or three kinds of modified business jets...and you put say twenty thousand tons of sulfuric acid into the stratosphere every year, and each year you have to put a little more and this doesn't, in the long run, mean that you can forget about cutting emissions. We will need to rein it in.
>
> **Colbert:** In the meantime, we're shrouding the earth in sulfuric acid.

Keith: So people are terrified about talking about this because they're scared that it will prevent us cutting emissions.

Colbert: Right, and also that it is sulfuric acid.

Keith: It is.

Colbert: Is there any possible way this could come back to bite us in the ass? Blanketing the earth in sulfuric acid…all over the earth?

Keith: Right question. But we put fifty million tons of sulfuric acid in the air now as pollution. It kills a million people a year worldwide.

Colbert: That's good or bad?

Keith: It's terrible.

Colbert: But it will be better if we put more in?

Keith: We're talking about one percent of that. A tiny fraction of that. So, we should reduce that sulfuric acid emission.

Colbert: But if it kills a million people—

Keith: It's bad.

Colbert: We only do one percent more, we're just killing ten thousand more people.

Keith: You can do math, okay! But that's—so killing people is not the objective here.

Colbert: Killing people is not the objective. I just wanted to be clear.

Keith: Actually, slowing climate change, actually stopping climate change in a way that could help people this generation, people living now. In a way, there's no other easy alternative.

While Colbert played the role of comedic provocateur, Keith played himself, revealing the plans for geoengineering are loaded with potential danger. Even though the math in their televised exchange was convoluted, to think that the proposed cure for climate change will kill people is bizarre, to say the least. I examined one research paper co-written by Keith that literally says the "falling" sulfuric acid "aerosol will add to the existing burden of near-surface fine particulate matter, degrading surface air quality and incurring public health damages in the form of increased respiratory disease mortality rates. By offsetting 1°C of atmospheric warming, greater concentrations…are formed from existing emissions, resulting in an additional 69,000 premature mortalities per year."[22]

Sixty-nine thousand deaths while trying to save the earth? What's to say that number is not an underestimate? And there are additional concerns: sulfates continue to gnaw into the ozone layer, the planet's primary solar radiation diffuser; and while SO_2 injected into the stratosphere could have a cooling effect in the

22   Sebastian D. Eastham, Debra K. Weisenstein, David W. Keith, Steven R. H. Barrett, "Quantifying the Impact of Sulfate Geoengineering on Mortality from Air Quality and UV-B Exposure," *Atmospheric Environment* 187 (2018): 424–434, https://keith.seas.harvard.edu/sites/hwpi.harvard.edu/files/tkg/files/eastham_ et_al._-_2018_-_quantifying_the_impact_of_sulfate_geoengineering_o. pdf?m=1530624251.

lower troposphere, it warms the stratosphere. We have no idea what kind of weather patterns that might stir up, because the models are incapable of predicting such things.

In an interview with CNBC, UCLA environmental law professor Edward Parson, an advocate of solar geoengineering, admitted spraying sulfur into the stratosphere will "mess with the ozone chemistry." Given the potential harm to both the environment and humans from sulfuric acid aerosols, Parsons was asked whether these known negatives are worth rolling the dice.

"Yes," he responded. "Respiratory illness is bad, absolutely. And spraying sulfur in the stratosphere would contribute in the bad direction to all of those effects. But you also have to ask, how much and relative to what?"[23]

And then there are those who see climate engineering as an easy way to milk the climate cult.

Luke Iseman and Andrew Song intend to take matters into their own hands and make a few bucks in the process. By putting a few grams of SO_2 into helium weather balloons and launching them into the upper atmosphere, the balloons will pop, release the sulfur dioxide, and bounce solar radiation back into space. Iseman and Song are actually making the sulfur dioxide themselves.

Oh, and they have a startup too: Make Sunsets.

They successfully raised $750,000 from venture capital firms to commercialize their stratospheric aerosol technology and have

23 Catherine Clifford, "White House Is Pushing Ahead Research to Cool Earth by Reflecting Back Sunlight," CNBC, October 13, 2022, https://www.cnbc.com/2022/10/13/what-is-solar-geoengineering-sunlight-reflection-risks-and-benefits.html.

launched a website offering people the chance to purchase "cooling credits" for ten bucks a pop. According to their website:

> We deploy our reflective clouds above 12.4 miles (20km) from the earth's surface using balloons. The reflective clouds stay up for about a year reflecting some of the Sun's rays just like the natural clouds below. Think of it as applying sunscreen spray to protect your skin from the sun. Just one gram of our clouds offsets the warming effect of one ton of CO_2 for a year.

Make Sunsets is not without its critics, even from fellow environmental activists. SilverLining (a progressive group that fits the definition of sustainable development) has lashed out, stating, "SilverLining strongly condemns Make Sunsets' rogue releases of material into the atmosphere and its efforts to market unsubstantiated 'cooling credits'. No one should be permitted to profit from releases into the environment with potential risks and no evidence of benefits."[24]

CONCLUSION

Geoengineering is loaded with unease. To think that bad actors could create unrelenting waves of devastating weather on an enemy could be like carrying out a prolonged military bombing raid at a fraction of the cost. And solar geoengineering? It is the hellish dreams of mad scientists and megalomaniacs.

---

24 "SilverLining Strongly Condemns Make Sunsets' Activities," Media Statement, SilverLining, February 13, 2023, https://www.silverlining.ngo/statement-make-sunsets.

Just as disturbing is the fact that no one is honest about any actual experiments that have been conducted with aerosols in the stratosphere without the knowledge or approval of the citizens of the countries where they are taking place. Circumstantial and other evidence is abundant that solar geoengineering has moved beyond the models. Would any reasonable person, when presented with the facts, believe that hundreds of scientists are spending their entire careers only on running computer simulations of stratospheric spraying? I highly doubt it.

There are over four million people in the United States with access to classified information, and over one million of them have a top security secret clearance. That's a lot of people in the know who, under penalty of the law, cannot speak on certain matters, most likely including geoengineering.

The challenges of governing such atmospheric capabilities are immense. A single government or cabal of governments, such as the United Nations, as well as wealthy individuals operating in the shadows of all governments, could carry out solar geoengineering, and every nation would be affected. Given the propensity for evil and lust for power among the elite, it is absolutely possible. How might these artificial alterations of weather patterns stunt plant photosynthesis, compromise wildlife habitat, and impact terrestrial ecological systems?

With the goals and focused determination by those types of people driving the green agenda, solar geoengineering is quickly becoming accepted by the millions of people in the climate change cult. Too many have had their minds transformed by the unending climate agenda messaging, to the point where they are willing to take a gamble. Holly Jean Buck, a professor at the

University at Buffalo, sums up the geoengineering mindset on behalf of the cult followers: "Can you not imagine someone in, say, 2050, who is suffering from extreme heat, wondering why their parents' generation decided to forbid research on something that might be able to cool the climate and save them from a dangerous heat wave?"

That is the propagandized climate change cult groupthink, and it is gaining more proponents at an alarming rate. This is a movement that must be stopped.

CHAPTER TEN

BRAVE NEW WORLD

OFTEN WHILE MY WIFE AND I were raising our four children, we would remind them of a simple scripture verse from the Psalms that was meant to temper their emotions whenever they were upset about something: "You can be angry, but don't sin."[1] That way they hopefully wouldn't end up doing something they might later regret.

The Hebrew word translated as "be angry" is *ragaz*, meaning "to be disturbed or agitated." The apostle Paul in his letter to the Ephesians quotes this same verse while giving instructions on how to better live as Christ did.[2] The takeaway for us: while there are sure to be experiences in life that will anger us, we need to guard our hearts so that we do not emotionally lash out in some harmful behavior.

I share this anecdote because as I've come to this closing chapter, I find myself angry, disturbed, and agitated, as if all the research that I've done for this book has come to a head and is about to burst within me. I find myself angry at the elite perpetrators of the environmental climate change agenda and their

---

1   Psalm 4:5 (Complete Jewish Bible).
2   Ephesians 4:26 (Complete Jewish Bible).

faux science. I'm disturbed that millions of people around the world have been sold an outright damnable lie about the climate. And I'm especially agitated by the lying media spokespersons and all the phony politicians, bureaucrats, lobbyists, activists, and social media influencers. My country is in the crosshairs of asymmetrical warfare intended to establish an ideology that has never ended well and never will.

However, I am reminded by the same verse I used to give my kids, rather than get angry and snap with vengeful curses, I can use my pen more mightily than I could a sword, because truth is from God and can never be extinguished. So now, I find myself yearning to make a measured and heartfelt final appeal to the reader.

DARTS IN THE DARK

Those who hawk the climate change agenda on the uniformed public are very much like deceitful door-to-door salesmen who throw darts in every direction with the expectancy of hitting a sucker wherever his projectiles land. Besides all their blather about a looming 1.5° increase in temperature, there's their impending year of doom: 2050. "If we continue on the current path," states the current National Oceanic and Atmospheric Administration website, in the United States, by 2050, "$106 billion worth of coastal property will likely be below sea level."[3]

Not only is that statement pure alarmist nonsense, NOAA seems to have positioned itself as an insurance risk assessor. If

---

3   "Climate Change Predictions," NOAA Office for Coastal Management, July 19, 2023, https://coast.noaa.gov/states/fast-facts/climate-change.html.

their prediction had any genuine merit, property insurance rates along coastal sectors of the US would be unfeasible.

But the predictions get progressively flagrant.

"Storm surges will have swept away coastal barriers erected at enormous cost and rising seas will have flooded the downtowns of major cities that once housed more than 100 million people," predicts *The Nation*. "When midnight strikes on New Year's Day of 2050, there will be little cause for celebration...it'll just be another day of adversity bordering on misery—a desperate struggle to find food, water, shelter, and safety."[4]

TIME magazine foresees the year 2050 experiencing "forest-fire season year-round, and the warmer ocean will keep hurricanes and typhoons boiling for months past the old norms."[5]

USA Today envisions the world in 2050 with "[m]ultiple plant and animal species...on the brink because of habitat change and destruction. Hotter, drier, and more erratic weather has hurt farming and farmers." They even envision fewer people eating meat because it will be "less popular and more expensive, given how much land and water it requires."[6]

These hyperbolic, fear-inducing predictions are designed to do one thing: force all citizens into conformity with a new world order. Predictions in *The Nation* article are followed by an appeal

4   Alfred McCoy, "Life Circa 2050 Will Be Bad. Really Bad," *The Nation*, December 20, 2021, https://www.thenation.com/article/environment/climate-future-disasters/.

5   Bill McKibben, "Hello from the Year 2050. We Avoided the Worst of Climate Change—But Everything Is Different," *TIME*, September 12, 2019, https://time.com/5669022/climate-change-2050/.

6   Elizabeth Weise, "On Earth Day, Scientists Tell Us What 2050 Could Be Like. Their Answers Might Surprise You," *USA Today*, April 21, 2023, https://www.usatoday.com/story/news/nation/2023/04/21/earth-day-what-us-look-like-2050/11607998002/.

for "new forms of global governance and cooperation," including a system where "a future U.N. could sanction in potentially meaningful ways a state that continued to release greenhouse gases into the atmosphere or refused to receive climate-change refugees."

To reference the previous chapter, if we have already reached a warming of 1.2°C since the beginning of the Industrial Revolution, what would another 0.3 degrees really look like? Would we even notice the difference in average temperature? Again, if we're willing to look, the observations reveal much. The average winter temperature at the North Pole is -40°C. Would -40.3 make a discernible difference? The average summer temperature in the Arctic is zero degrees. Will 0.3°C actually cause all the ice to melt?

And what about Greenland with its summer temperature averaging 10°C (50°F)? Is 10.3°C going to cause the glaciers to liquefy?

Well, that is what we're told is going to happen in all these scenarios. Yahoo! News even reported that when the glaciers melt, the oceans will rise twenty-three feet and, "Once we start sliding, we will fall off this cliff and cannot climb back up."[7]

These are extremely fraudulent forecasts directly reaching back to the failed alarmist predictions of von Liebig, Schneider, Ehrlich, and Carson.

The fact is that since the Ice Age, global sea levels have been gradually increasing as the accumulated ice and snow from that bitterly cold event continue to melt and trickle into our great

---

7   Rob Waugh, "Greenland Ice Sheet Which Could Raise Sea Level by 7m 'Nears Point of No Return,'" Yahoo! News, March 29, 2023, https://uk.news.yahoo.com/greenland-ice-sheet-which-could-raise-sea-level-by-7m-nears-point-of-no-return-152821651.html.

oceans and seas. According to the United Nations' International Panel on Climate Change (IPCC) 2007 Fourth Assessment publication, over the past twenty thousand years, sea levels have increased nearly four hundred feet.[8] But over the past century, the average sea level has risen a mere 1.8 millimeters per year.[9] How can that amount be understood? By placing your thumb and forefinger as close as possible without touching. That's the 1.8-millimeter rise in sea levels each year for the last hundred years. Hardly warranting the tsunami sirens that are sounding, and it has nothing to do with fossil fuels or carbon dioxide.

One last practical temperature tidbit: Phoenix, Arizona, the hottest and fastest-growing big city in America with an average summertime high of 100°F, would *not* run out of air conditioning due to a 0.3°C temperature increase, unless their energy grid goes absolute net zero and totally runs off renewable energies.

But as the climate cult leaders cleverly changed the brand of their crusade from "global warming" to "climate change," they quickly throw another dart in the dark and claim we can't just focus our gaze on temperature only, because those menacing warming temps are causing weather worldwide to go amok, most often in the supposed rise of the biggest storms on Earth.

BIGGER STORMS?

National Public Radio (NPR) reports, "There is a growing body of evidence showing that hurricanes are intensifying more

---

8   International Panel on Climate Change, Fourth Assessment Report, Chapter 5, 2007, 409.

9   Ibid., Chapter 10, 821.

quickly, turning from less-serious storms to very strong ones in hours or days."[10]

More fraud. NPR is likely referring to something known as *extratropical transition*, which involves a myriad of atmospheric features, but to say the storms are undergoing unusual intensification is reckless.

Dr. William Gray, referenced in chapter three, was unquestionably the world's foremost hurricane forecaster, founding the Tropical Meteorology Project at Colorado State University in the 1960s, where he developed the art of forecasting hurricanes in the Atlantic Basin, including the Gulf of Mexico. I interviewed Dr. Gray on my radio show numerous times, and numerous times he told my radio audience, "I am of the opinion that global warming is one of the greatest hoaxes ever perpetrated on the American people."

As a foremost expert, Dr. Gray's opinion was founded on real science, not based on gut feeling. "All my colleagues that have been around a long time—I think if you go to ask the last four or five directors of the National Hurricane Center—we all don't think this is human-induced global warming," said Dr. Gray.[11]

Indeed, another pioneer in hurricane research, the thirteen-year director of the National Hurricane Center, Dr. Neil Frank, presented his opinions to me regarding the 2020 hurricane season, which experienced thirty-one such storms in the Atlantic Ocean. Activists immediately claimed the storm count was the

10   Alejandra Borunda, "What's the Connection between Climate Change and Hurricanes?" NPR KQED, August 30, 2023, https://www.npr.org/2023/08/30/1196865225/whats-the-connection-between-climate-change-and-hurricanes.

11   Interview with Dr. William Gray on KSFO, San Francisco, April 7, 2007.

result of climate change, but Dr. Frank strongly disagrees. In a December 2023 email exchange, Dr. Frank made it truly clear that even though 2020 was an active season, it had nothing to do with anthropogenic atmospheric changes. "Remember," he reminded me, "prior to the use of weather satellites in the 1970s hurricanes were noted by way of sightings by ships at sea." Once satellite observations were employed, the number of hurricanes increased greatly.

He also explained that over the past 150 years there have been three distinct multi-decadal hurricane escalations. "The first one," he said, "occurred in the late 1800s as the earth emerged from the little ice age and there was a flurry of hurricanes. In 1893 a strong hurricane reached the Georgia and South Carolina coasts, a 15–20 ft. storm surge inundated the coastal islands. Though population was a small fraction of todays, between 2,000 and 3,000 died, making that the second deadliest hurricane in U.S. history. The same year another major hurricane killed 2,000 in Louisiana. All together five hurricanes hit the U.S. in 1893, something that's happened only 4 times in over 150 years (1886, 1893, 1916, 1933)—all long before CO_2 levels rose enough to theoretically cause rapid global warming. As a matter of fact, the most active year for land-falling hurricanes in the lower 48 was in 1886 when 7 hurricanes crossed the Gulf Coast."

Dr. Frank described the second wave, which came about many years later. "Florida was hit by 7 major hurricanes in the 1940s and in the 1950s 6 major hurricanes hit the Carolinas." As for the recent increase in annual hurricanes, Dr. Frank explained it followed a nearly thirty-year lull. "The high number of storms in 2020 was because we are [now] at the peak of a very active hurricane period."

Dr. Frank's comments align well with a quote he made to the *Washington Post* in 2006 interview on climate change: "It's a hoax," he plainly said."[12] As for tornadoes, they too cannot be tied to the use of fossil fuels. A recent peer-reviewed research paper reveals that both tornado frequency and intensity have *decreased* since 1950.[13]

BOILING OCEANS? SINKING ISLANDS?

Since Al Gore was the first to utilize fear appeal with the term "boiling oceans," boiling has become the go-to adjective for all the climate agenda's usual spokesmen, like the UN secretary-general who claimed that we have entered the "era of global boiling."

The boiling temperature for water is 212°F. Oceans are not boiling.

In a July 26, 2023, Associated Press news story about a temperature buoy off the coast of Florida recording a water temperature of 101.1°F, the reporter alarmingly claimed, "The water temperature around the tip of Florida has hit triple digits—hot tub levels—two days in a row. Meteorologists say it could be the hottest seawater ever measured...."[14]

Let's apply some common sense here: a high single seawater temperature reading on a single day is not an indicator of climate change, nor is it unprecedented. The buoy is located in Manatee Bay in very shallow water, in fact, so shallow that it's dark and

12 Joel Achenbach. "The Tempest," *Washington Post,* Sunday, May 28, 2006.

13 Jinhui Zhang et al., "Time Trends in Losses from Major Tornadoes in the United States," *Weather and Climate Extremes,* vol. 41, September 2023, https://www.sciencedirect.com/science/article/pii/S2212094723000324?utm_source=substack&utm_medium=email

14 Seth Borenstein, "South Florida Waters Hit Hot Tub Level and May Have Set World Record for Warmest Seawater Ever," Associated Press, July 26, 2023.

stagnant, and at low tide, the buoy rests atop the bottom muck. The July 25 record occurred during a low tide, so, obviously, the buoy temperature is not representative of Atlantic Ocean temperatures.

And, as an aside, a one-hundred-degree hot tub temperature is actually tepid.

But on to rising sea levels.

In 2018, one of the all-time sea level experts, Stockholm University professor Nils-Axel Mörner, who served as an expert reviewer for the UN's International Panel on Climate Change (IPCC), didn't pull any punches illustrating the blatant bias of many in the research business:

> I have been researching sea-level changes my entire life, traveling to 59 countries. Hardly any other researcher has so much experience in this field. However, the IPCC has always misrepresented the facts on this topic. It exaggerates the risks of a sea-level rise enormously.
>
> We were able to prove that the sea level in Fiji from 1550 to about 1700 was about seventy centimeters higher than it is today. Then it sank and was about fifty centimeters lower in the 18th century than it is today. Then it rose to about the current level. In the last 200 years, the level has not changed significantly. For the past 50 to 70 years, it has been stable.[15]

---

15 "Nils-Axel Mörner: 'These Researchers Have a Political Agenda,'" *Basler Zeitung*, February 18, 2018, https://breakingviewsnz.blogspot.com/2018/02/gwpf-newsletter-jeremy-corbyn-promises.html.

Why do so many climate "experts" warn about sinking islands? Because they have a political agenda. As Dr. Mörner explained, the IPCC was founded with the purpose of *proving* man-made climate change. He even revealed that the panel's initial reports declaring dangerously rising sea levels had "twenty-two authors, but none of them—*none*—were sea-level specialists."[16]

INCH BY INCH

No matter the facts, the agenda's influencers present scenarios that invoke a natural sense of alarm and then, without missing a beat, they provide the remedies: sustainable development, social justice, and social equity. Recall the words from the United Nations 1987 declaration, *Our Common Future*:

Meeting the basic needs of all...

The opportunity to fulfill aspirations for a better life...

A new era of economic growth for the nations...

Those poor get their fair share...

Adopt lifestyles within the planet's ecological means...

The same document declares, "Sustainable global development requires that those who are more affluent adopt life-styles within the planet's ecological means—in their use of energy, for example."[17]

It also demonizes fossil fuels, insisting, "With the exception of CO_2, air pollutants can be removed from fossil fuel combustion processes at costs usually below the costs of damage caused

16 "Sea Level Expert: 'It's Not Rising!,'" *21st Century Science & Technology*, Fall 2007.
17 *Our Common Future*, subchapter 3.29.

by pollution. However, the risks of global warming make heavy future reliance upon fossil fuels problematic."[18]

Quite obviously, the plans of the climate change agenda are extensive, particularly the most recent schemes being pushed in the United States, including those introduced by congressional Democrats and the Biden administration promoting net zero and the Green New Deal (GND), both designed to formally take a sledgehammer to Lady Liberty.

The GND was originally articulated in a 2007 *New York Times* article by Thomas Friedman. The label harkens back to the set of reforms and public works projects undertaken by President Franklin Roosevelt in 1933–1935 in response to the Great Depression. Friedman wrote that a Green New Deal would create a comprehensive solution to climate change while promoting economic growth through the development of renewables. His popular article was followed by a book further promoting the concept, and soon the plan was endorsed by the United Nations who, as you have discovered, had been pining for such a plan for decades.

The GND was formally introduced in 2019 by two extremely progressive lawmakers, US representative Ocasio-Cortez, D-NY, and US senator Ed Markey, D-MA. Though it has yet to be passed into law, a lighter version, deceptively titled the Inflation Reduction Act (IRA), has been. Heralded by Joe Biden's Environmental Protection Agency as "the most significant climate legislation in U.S. history,"[19] the IRA "invests an

---

18 Ibid., Chapter 7, subchapter 2.18.
19 "Summary of Inflation Reduction Act Provisions Related to Renewable Energy," Environmental Protection Agency, updated June 1, 2023, https://www.epa.gov/green-power-markets/summary-inflation-reduction-act-provisions-related-renewable-energy.

unprecedented $369 billion in climate action"[20] and represents a perfect example of the Marxists' creeping inch-by-inch style that they rely on so heavily to accomplish their political goals.

The IRA, along with the previously enacted Infrastructure Investment and Jobs Act, allows the Democrats to institute key elements of net zero—which seeks to drastically reduce the use of fossil fuels by 2050—and dole out hundreds of billions of dollars in subsidies to renewable energy developers, in-the-know investors, and special-interest stakeholders.

Regardless of the fact that before passage of the IRA, US emissions already had been falling for decades, driven largely by market forces that included the increased penetration of natural gas allowed by improved extraction technology,[21] a growing share of global carbon dioxide emissions by developing nations continued to rise: China's dirty coal-based grid contributed about one-quarter of the world's emissions in 2021,[22] while the US accounted for about 12 percent, down from almost 17 percent in 1990.[23]

The stark reality is a world without fossil fuel is simply untenable, as fossil fuel supplies nearly 85 percent of the world's energy consumption. America's oil and natural gas industry accounts for more than ten million jobs. The US has both the resources and

20 Mikyla Reta, "How the Climate Bill Could Boost the Justice40 Initiative," National Resource Defense Council, September 19, 2022, https://www.nrdc.org/bio/mikyla-reta/how-climate-bill-could-boost-justice40-initiative.

21 Nicolas Loris, "Pursuing Policies to Drive Economic Growth and Reduce Emissions," Heritage Foundation Backgrounder No. 3444, October 16, 2019, https://www.heritage.org/energy-economics/report/pursuing-policies-drive-economic-growth-and-reduce-emissions.

22 Hannah Ritchie, Max Roser, and Pablo Rosado, "CO_2 and Greenhouse Gas Emissions," Our World in Data, based on Climate Analysis Indicators Tool, August 2020, https://ourworldindata.org/co2-and-greenhouse-gas-emissions (accessed June 6, 2023).

23 Ibid.

technology to be the planet's number one producer of natural gas and crude oil as well as a major exporter of energy. And, as opposed to other countries, the US is second to none in fossil fuel extraction with the utmost respect for the environment.

For China, however, net zero is the goose that laid the golden egg. Net zero would increase US reliance on communist China to produce rare earth minerals required for electric vehicles and their batteries. China is also the dominant producer of solar panels and wind turbines, with a 70 percent market share. Meanwhile, Chinese president Xi Jinping pledges that his country will achieve net zero by 2060.

NET ZERO MADNESS

The vanguards of net zero, though, never speak of the useful everyday products that would virtually disappear if petroleum products were banished. Our favorite pillows and mattresses? Made from fossil-based petrochemicals—as are most of our wardrobes that contain nylon, polyester, rayon, and other materials. And then there are plastics used to make all manner of containers, from aspirin bottles to toothpaste tubes, to cosmetic and hygiene products, to milk and food packages, to trash cans and storage bins. Much of your furniture, car, tools, and toys, too, consist of varying amounts of plastic, which has been demonized as indecomposable in landfills and would remain in that state for a thousand years. I don't know which unfounded myth is more absurd, that one or the one about the supposed gigantic *island* of floating plastic somewhere in the Pacific Ocean.[24]

---

24 *Brian Sussman Show* podcast (audio episode 210, video episode 7), interview with plastics expert Dr. Chris DeArmitt.

Then there is the roof over your head. Asphalt shingles and tar paper used in roofing both consist of materials involving the release of carbon dioxide during their production. A metal roof would be out of the question, too, as steel is made from iron ore, which is iron oxide. To get the iron out of the oxide, steel plants use coking coal, which produces carbon monoxide. The carbon monoxide cleaves the oxygen off the iron, a chemical process that requires hydrocarbons. The US manufactures over ninety million tons of steel a year out of two billion tons produced worldwide. Cement production is not compatible with net zero either, as it accounts for 8 percent of global CO_2 emissions.[25]

Again, it is absurdly obvious that wind and solar energy could never power a manufacturing plant for any of these vital products.

POLICY CHANGES ASAP

There is one problem Americans should never experience: paying high gasoline and diesel prices at the pump or high utility bills to heat or cool your home. The United States of America is one of the most energy-rich nations on planet Earth. As a businessman, President Donald Trump recognized that leftist lawmakers have been foolishly chipping away at the energy sector for decades. Once in office, Trump used his executive authority to unleash energy independence by taking the federal shackles off oil and gas exploration, drilling, fracking, refining, and delivery. In fact, it took forty-three years—from 1977 to 2020—for a brand-new

---

25 Jocelyn Timperley, "If the Cement Industry Were a Country, It Would Be the Third Largest Emitter in the World," *Carbon Brief*, September 13, 2018, https://www.carbonbrief.org/qa-why-cement-emissions-matter-for-climate-change/.

oil refinery to be constructed in the US, but it happened because of policies instituted during the Trump administration.

However, all that radically flipped on President Biden's first day in office when, with the stroke of a pen, he cancelled the permits for completion of the Keystone XL oil pipeline. Besides stopping a safe river of crude oil from securely flowing to US refineries, Biden's order killed thousands of good-paying jobs. One Keystone contractor, TC Energy, had deals with four labor unions to hire forty-two thousand employees.[26] This sent a powerful signal that the Biden administration and congressional Democrats would make it harder for American energy producers, refiners, and workers to meet the needs of the American people. With Biden at the helm, the environmental lobbyists have a powerful partner who is granting all of their wishes to squash our fossil fuel sector and expand inconstant renewables. Adding a further insult to the American people, on June 20, 2022, while vacationing in Delaware, President Biden was asked by reporters about his energy policies, arrogantly stating, "My mother had an expression: out of everything lousy, something good will happen. We have a chance to make a fundamental turn toward renewable energy, electric vehicles, and across the board."[27]

If a Republican wins the White House in 2024, and the GOP is able to win both the House and Senate, plans must be implemented immediately across all federal agencies to deconstruct decades of damage done to our nation in the name of climate change and the environment. Barriers that limit oil and gas

---

26 Thomas Catenacci, "Biden Admin Quietly Admits Canceling Keystone XL Pipeline Cost Thousands of Jobs, Billions of Dollars," Fox News, January 5, 2023.

27 Kristine Parks, "Backlash Ensues as President Biden Suggests Inflation a 'Chance' to Make 'Fundamental Turn' to Clean Energy," Fox News, June 20, 2022.

supplies must be removed, pipeline capacities must be increased, and stifling regulations eliminated. This would easily bring the price of energy down and have a huge positive effect on reducing inflation.

Eliminating the federal Office of Energy Efficiency and Renewable Energy—whose mission is to "transition America to net-zero"[28]—must also be a priority in Washington, especially since taxpayer dollars shouldn't be used to subsidize preferred businesses and energy resources anyway, because these financial offerings distort the market and undermine energy reliability.

Solar and wind are not the solution either. Instead, a reliable grid and security should be the focus. Likewise, the Office of Clean Energy Demonstrations (another bureaucracy dedicated to transitioning the country to a decarbonized energy system) should be shuttered.

We also must insist that our elected representatives repeal, or gut, the IRA and the Infrastructure Investment and Jobs Act, which are doling out hundreds of billions of dollars in subsidies to renewable energy developers, investors, and stakeholders. The focus should be ensuring that customers have affordable and reliable electricity, natural gas, and oil. Democrat lawmakers sold the IRA to the American public on a lie, claiming it was a federal deficit reduction package. Taking an axe to our massive budget deficit and unfathomable national debt (which at this writing is approaching $34 trillion) cannot occur without reducing federal spending, which in recent fiscal years consumed an aver-

---

28 "About the Office of Energy Efficiency and Renewable Energy," Department of Energy, https://www.energy.gov/eere/about-office-energy-efficiency-and-renewable-energy.

age of almost 29 percent of gross domestic product (GDP).[29] That figure is higher than any other year on record outside of World War II.

Habitual overspending by career politicians who allow legislative bills to be authored by lobbyists is the root cause of America's soaring deficits. Tax revenues already hit historic highs even before the IRA was passed and cannot rise higher without inflicting serious damage to American households and US businesses. To avoid a future fiscal collapse, Congress must pass meaningful reforms to reduce both mandatory and discretionary spending as well as stop the proliferation of stakeholder capitalism, whereby environmental groups and radical activists can exert control over businesses by directly strong-arming corporate boards to improve their environmental, social, and governance (ESG) ratings. The Securities and Exchange Commission (SEC) is also accommodating the ESG movement by pushing expensive mandates for climate change disclosures from public companies, an effort abetted by the Department of Labor's recent regulation about fiduciary investing. All of this must end.

STOPPING THE GREEN NEW DEAL

We also must enlighten those who have been unwittingly drawn into the cult and educate those whose personal antennae tell them something is really wrong about the Left's cherished Green New Deal and the Biden-Harris IRA. It's vital America understands the massive amounts of money required to fund these pipe

---

29 Office of Management and Budget, "Table 1.2: Summary of Receipts, Outlays, and Surpluses or Deficits (-) as Percentages of GDP: 1930–2028," https://www.whitehouse.gov/omb/budget/historical-tables/ (accessed February 9, 2023).

dreams without delivering any tangible environmental benefits, all the while further centralizing power and control over the daily lives of Americans while providing handouts to lobbyists and stakeholders.

Here are just four items I believe reveal the complete insanity of these plots.

SOCIAL JUSTICE AND EQUITY

Illustrating the intersection of sustainable development, social justice, and social equity, the Green New Deal requires billions of dollars to fund "environmental justice" programs in higher education. According to the resolution, it would provide, "resources, training, and high-quality education, including higher education, to all people of the United States, with a focus on frontline and vulnerable communities, so that all people of the United States may be full and equal participants in the Green New Deal mobilization."[30] This is a plan to radicalize students and create the next generation of political activists while rewarding otherwise unemployable environmental instructors.

CLIMATE CORPS

This frightening idea has been implemented by the Biden administration with the announcement that twenty thousand green-shirted youths are being hired into the new American Climate Corps. According to a tweet from Joe Biden's POTUS account, the program will train them to be propagandists for "clean energy, conservation, and climate resilience related skills."[31]

---

30 Green New Deal Resolution, 11.
31 President Joe Biden, X (Twitter), September 20, 2023.

The GND is smattered with multiple "civilian climate corps" programs. Just the name, "climate corps," illustrates the ideology behind this decision. Throughout history, authoritarian regimes have always found it useful to recruit their youths. The GND would earmark billions to fund well-paying federal jobs at parks and forests, and many so-called "conservation" projects designed to benefit non-profit environmental organizations.

JUSTICE40 INITIATIVE

The Justice40 Initiative is a Biden White House plan directing 40 percent of all climate and clean energy financial outlays to "disadvantaged communities, identified by the Climate and Economic Justice Screening Tool." This screening apparatus was put forth by Executive Order 14,008, demanding that "communities that are disadvantaged because they are overburdened and underserved" be identified.[32] Despite the fact that Justice40 will make available $60 billion to community stakeholders, the powerful environmental legal advocacy group, the National Resources Defense Council, has "expressed frustrations with some of the bill's provisions,"[33] likely because the pool of money is not, from the perspective of their three million members, large enough.

ELECTRIC VEHICLES

Another wish from the GND that has made it into the IRA includes $43 billion of free money to subsidize the purchase of

---

32 About Page, Climate and Economic Justice Screen Tool, https://screeningtool. geoplatform.gov/en/about.

33 Mikyla Reta, "How the Climate Bill Could Boost the Justice40 Initiative," National Resources Defense Council, September 19, 2022, https://www.nrdc. org/bio/mikyla-reta/how-climate-bill-could-boost-justice40-initiative.

electric vehicles and new home appliances.[34] New plug-in cars and trucks qualify for federal tax credits worth up to $7,500, while used vehicles can receive up to $4,000. The tax credits can also be used as a point-of-sale discount on the price of new and used electric vehicles, essentially allowing a buyer to transfer their tax credit directly to a car dealer. Thus, the dealer, who must register with the US Department of the Treasury, gets an advance payment taken from the buyers' tax credit. Most consumers are using this free money as a down payment. As you discovered previously, taxpayer subsidies are the carrots being dangled to lure consumers into the plug-in world of otherwise very expensive transportation.

Additionally, if you like your natural gas stove or fireplace, the IRA is part of a broader effort to make these appliances relics of the past by spending billions to promote electric appliances and entice consumers with federal tax credits and deductions to ditch the gas appliances for new all-electric ones. This is a recipe for disaster. When the Texas electric grid failed during an ice storm in 2021, hundreds of residents died of hypothermia, while millions of residents were able to keep slightly warm by using their gas fireplaces and cook food with their gas ranges and ovens. Residents with gas ranges also boiled water to assure its purity, something they could not have accomplished on an inoperative electric cooktop.

---

34 "The Inflation Reduction Act: Here's What's in It," McKinsey & Company, October 24, 2022, https://www.mckinsey.com/industries/public-sector/our-insights/the-inflation-reduction-act-heres-whats-in-it.

THE EPA SLUSH FUND

The Environmental Protection Agency (EPA) is provided with over $135 billion each year to spend and distribute—that's more than plenty. However, the IRA has created a $27 billion slush fund many on Capitol Hill have nicknamed the Green Bank. The fund was designed to set up two pots of money, $20 billion for the General Assistance and Low-Income and Disadvantaged Communities Program and $7 billion for the Zero-Emissions Technologies program. All of this taxpayer largess would be given to non-profits. The grants must align with the Justice40 Initiative.

TOTAL COST

Financial estimates of the proposed Green New Deal are all over the map. Republican lawmakers have consistently priced it at $93 trillion, while leftists like Noah Kaufman, a former researcher with the Obama administration and current scholar at Columbia University's Center on Global Energy Policy, says the Green New Deal will cost "basically nothing."[35] Meanwhile, the cost for reaching the Biden administration's goal of net zero carbon emissions by 2050 will be $275 trillion, according to a report from McKinsey & Company.[36]

---

35 Deborah D'Souza, "Understanding the Green New Deal & What's in the Climate Proposal," *Investopedia*, May 28, 2022.

36 Press Release, "The Cost Will Not Be Net Zero," McKinsey & Company, February 18, 2022, https://www.mckinsey.com/featured-insights/sustainable-inclusive-growth/chart-of-the-day/the-cost-will-not-be-net-zero.

THE USELESS CLASS

"The causes of the environmental crisis are not mysterious. Solving these problems does not require more science but more socialism," says the progressive organization openDemocracy, an organization based in the United Kingdom and funded by the Soros Open Society Foundations, the Rockefeller Foundation, the Packard Foundation, and the Ford Foundation.[37]

Those orchestrating the climate change agenda hate the ideals that made America great—life, liberty, and the pursuit of happiness. Taking a wrecking ball to the economy, doling out electricity like a miser, reducing the population, and limiting personal financial opportunity is the climate change cult's plan to keep the "lesser-minded" under their control. The openDemocracy article clearly spells it out:

> It should not be surprising that socialising the economy means the politicisation of everything. Instead of interacting as autonomous consumers, we have to negotiate amongst ourselves as to whether to have wasteful levels of energy consumption paired with perilous solar geoengineering, or renewable infrastructure teamed with energy quotas that help the poor even as they constrain the rich.

The summation of those two sentences can be found in two phrases: "help the poor" and "constrain the rich." This represents

---

37 Troy Vettese and Drew Pendergrass, "A Planned Economy Is the Way to Save the Planet," openDemocracy, April 26, 2022, https://www.opendemocracy.net/en/oureconomy/planned-economy-needed-to-save-the-planet/.

the full-scale application of social justice and equity carried out with complete government sanctioning. Those who refuse to comply with their climate agenda will eventually be shut out from participating in the economy—their ESG ratings will be pathetically low—and consequently, from society altogether.

We already are under constant surveillance by Big Tech, with our phones, computers, Alexas, ring cameras, and the internet monitoring everything we do online and offline. Everywhere we are and everywhere we go, we're being tracked. Our relationships, finances, purchases, health issues, religious beliefs, and politics are a constant stream of data being monetized by commercial interests, shared with governments, and often stolen by hackers.

Post a selfie while skiing in Colorado with the caption, "Loving this global warming—lol," and the ESG compliance department at the firm managing your 401K decides to drop you. Or that end-of-the-year government health insurance rebate you had hoped to receive when you downloaded a simple app on your phone and linked it to your watch comes back to bite you because the app monitors your vital health statistics, including heart rate and blood pressure 24/7, and you're deemed not healthy enough for that rebate. Or maybe Big Tech notices that your blood pressure rises when you're complaining about the increasingly woke culture while they are monitoring your conversations. Then HR finds out, and you're fired for ruining the company's ESG standing. And welcome to the "useless class." We are closer to this than most realize.

"Useless class" is a term often used by Yuval Noah Harari, a WEF advisor who co-authored *COVID-19: The Great Reset* with Klaus Schwab. Though Harari applies the phrase when speaking of those whose careers have been left behind because of future advancements in artificial intelligence (AI), I think it is a safe bet to say that given the fact that climate change is a top WEF issue, the useless class will come to include those who cherish liberty as well.

Harari perceives an eventual world with three class strata: the elite, the workers, and the useless class. "Those who fail in the struggle against irrelevance would constitute a new 'useless class,'" he explains without apology. His evil assessment is that they are useless from the standpoint "of the economic and political system. And this useless class will be separated by an ever-growing gap from the evermore powerful elite."[38]

From what I've heard and read of him, Harari never mentions those with mental, emotional, or physical handicaps, but it doesn't tax one's imagination much to figure out how they will fare in his twisted universe. Undoubtedly, they will likely be counted with the lazy, unreliable, and "Give me liberty, or give me death" class.

Harari explains:

> Again, I think the biggest question in maybe in economics and politics of the coming decades will be what to do with all these useless people?
>
> The problem is more boredom and how what to do with them and how will they find some sense of meaning in life, when they are basically meaningless, worthless?

38 "Read Yuval Harari's Blistering Warning to Davos in Full," *Davos Agenda*, World Economic Forum, January 24, 2020, https://www.weforum.org/agenda/2020/01/yuval-hararis-warning-davos-speech-future-predications/.

My best guess, at present is a combination of drugs and computer games as a solution for [most]. It's already happening. Under different titles, different headings, you see more and more people spending more and more time or solving their inner problems with the drugs and computer games, both legal drugs and illegal drugs.[39]

CONCLUSION

The climate change agenda and its propositions are not about a post-fossil fuel world, nor are they about a utopia where everyone gets along. The elite power brokers of this evil agenda only want absolute control of everyone and everything. They will feign support for democracy as long as everyone votes for their climate change politicians and supports their climate change policies that pursue sustainable development. But oppose their ideas and goals, and the climate change elites will extort your God-given right to life, liberty, and the pursuit of happiness and demonize you as a sinful, energy-consuming cancer that can only be cured by being reprogrammed into their brave new world—or eradicated.

Their heinous objective is nothing short of a total assault on our nation's history and values and demands nothing short of our total opposition and resistance to stop it.

For too long, Marxists have used our schools, entertainment industries, and news media to indoctrinate the developed

---

39   Yuval Noah Harari, "How Drugs & Video Games Have Been Instrumental in Controlling the Population," World Thought Leaders, YouTube, https://www.youtube.com/watch?v=VZP5lIzGNT8.

nations—most specifically, the United States—for at least three generations. But when Marxism gains power, it does not tolerate other world views, nor, by its nature, can it. It always demands that opposing opinions be censored, to the point of driving those who express such positions from the public square, and as history has proven, things can turn ugly very quickly against those who do oppose it. But as Marx stated, "History means nothing." So, consequently, the new Marxists continue playing for keeps, just like the old ones did, and damn those who get in their way.

Aldous Huxley's universal slogan in *Brave New World* is, "Everyone belongs to everyone else."

For those caught in the web of the climate change agenda, that slogan is their hope and dream. But in reality, if the agenda is allowed to succeed, everyone will be subject to everyone else—in the elite class—and global tyranny will reign.

In works of science fiction, the world is often presented as being scientifically logical; in fantasy novels and films, it is mostly depicted as supernatural. But the real-life climate change world is scientifically *illogical* and *demonically* supernatural. The climate change cult and those who control it are *not* a conspiracy theory. They are part of an open, avowed, well-planned scheme that may read like fantasy, but is destructive to every human being in the entire world. Their marketing slogan spells out exactly what their goal is: "You'll own nothing and you'll be happy."[40]

So, now maybe you understand my anger expressed on the opening page of this chapter a little better. And if you find yourself feeling the same way, you can join me in heeding my own

---

40 "You'll own nothing and you'll be happy" (alternatively, "You'll own nothing and be happy") is a phrase originated by Danish politician Ida Auken in a 2016 essay for the World Economic Forum.

advice, "Be angry, but don't sin." Make your voice heard, petition your elected officials, vote the bad guys out and good guys in, highlight and underscore what you have discovered in this book, review the footnotes for yourself, use this information as a tool to steer people away from the cult, stay on top of your local school board and consider running yourself.

Just, please, do not remain silent about what you have read and learned.

AFTERWORD

THERE'S A BOB DYLAN SONG that describes a snapshot from hell, where we find Sigmund Freud and Karl Marx receiving their due. The lyrics to "My Own Version of You," written in 2020, are worth noting.[1]

Freud's influence in the field of psychology and human subconsciousness has been prominent for more than a hundred years, especially among broadminded thinkers and academicians who perceive themselves as intellectually elite.

Like so much of the propaganda from the climate agenda we have examined in this book, Freud's most notable ideas were not rooted in truth. A prime example is his Oedipus complex, which holds that every young boy dreams of having sex with his mother and murdering his father, whom the boy perceives as a rival. However, because the boy also possesses the foresight to realize that his father is his protector, the child represses his homicidal tendency.

The Oedipus hypothesis is madness, doubling as a slap in the face to those who believe the biblical adage of *imago dei*, or "made in His image."

We *are* made in God's image, and, because of that, each human being has the potential to bring wonderful goodness

---

1 Bob Dylan, "My Own Version of You," from the album *Rough and Rowdy Ways*, Columbia Records, 2020.

to society at large. Freud's teaching, though, denies such inherent virtue.

And then we have Marx.

As you have seen, Marx certainly never perceived humankind as *imago dei*. Instead, he went to his deathbed contending that a privileged few are born with a significant transcendent intelligence giving them the ability—and authority—to organize the masses into an easily controllable group. Like Freud, Marx's philosophy resonates with the irreligious and the *haut monde*, or the high and fashionable society.

As I interpret Dylan's song, I sense he is reminding us that throughout the long slog of human history, Marx's reasoning has proven itself to be forcibly destructive.

This is not the first song in which Dylan has taken aim at the co-author of *The Communist Manifesto*. On his 1979 *Slow Train Coming* album, there is a persuasive song entitled "When You Gonna Wake Up?" It's a politically charged composition that prophetically describes the steady societal breakdown that is culminating today. He speaks of Marx squeezing life from the throats of the masses, Deep State actors tying minds into knots, and phony philosophies further polluting our thoughts.

Dylan's observations are presciently on the money.

Observing society through the lens of my own lifetime, I naturally tend to become very pessimistic about the direction we are headed. However, just like the lessons I have learned through the observable changes in climate, I must look beyond my fleeting time on this earth and realize it is the ebb and flow of society over millennia that gives us the most accurate picture of history rather than over just a few decades or even a century or two. Throughout the past, there have been horrible wars and cataclysmic natural disasters. Regions of the world have experienced great repression, famine, disease, and cultural decay. But the pages of history always reveal a ray of veracity that somehow shined through the darkness like a beacon, and a new era of hope and prosperity was born and began to blossom.

Indeed, such was the very case in the days of King George of Great Britain when he ruled his vast empire with a cruel and brutal hand. Those who ran afoul of his edicts were imprisoned without trial and often hanged or burned at the stake. Yet something unique, a tiny flicker of light, was stirring to the west across the Atlantic Ocean in colonial America. Even though British immigrants to the new land were quite divided on many issues, the more they experienced the blatant disregard and belligerence of their king, the more a yearning for freedom and self-autonomy was rising in their souls.

In the southern colonies, their slogan had become, "Damn your soul, grow tobacco," because slavery in the tobacco trade, and at large, was not only tolerated in the South, but strongly encouraged. While in the North, the Christian faith of the Puritans and Pilgrims yielded great positive influence among the populace, despite King George's callousness for his subjects as he aimed to make their lives miserable and desperate. But even with those challenges, the conception of the greatest nation in the history of the world was taking place. Little did anyone realize a country known as the United States of America would become the ray of truth and promise to millions trapped in darkness across the globe.

It was a profound spiritual "Great Awakening" that sparked that tiny ray in the 1730s. It began in New England and then spread south over the next two decades and witnessed countless colonists suddenly discovering inspiration and encouragement through personal faith in Jesus Christ. This spiritual enlightenment ignited courage in the people to stand up to Great Britain and demand independence. Their unflinching resolve led to the proclamation of a Declaration of Independence, which, in turn, led to a great and costly war. Many patriots lost their lives, fortunes, and sacred honor, but freedom was achieved, paving the way to a constitution anchored in the recognition of the inalienable rights granted by God, not by other men.

America still had to endure further pain over the issue of slavery, and it took a Civil War and more than six hundred thousand deaths to finally put an end to the evil practice of humans owning other humans. Though the scars of slavery remained, some even to this day, America moved steadily forward and began to flourish, until it became the most prosperous and benevolent nation in the history of the world.

By the late nineteenth century, the deadly seeds of Marxism and varying degrees of authoritarianism were being sown around the world, seeds that unleashed a twentieth century filled with Marxism, Leninism, Stalinism, Nazism, Imperialism, Maoism, and radical Islam. Countless multitudes brutally perished because of the evil teachings of those doctrines. And although Lenin, Stalin, Hitler, and even Osama bin Laden were eventually killed, many bad and worse actors have risen to take their place, including those who patiently continue to undermine America's great ethos of life, liberty, and happiness.

Marxism, without any doubt, has taken firm root in America, and, as you have discovered, one of the many weapons being used to pummel us into its submission is the climate agenda. I pray that after reading this book, at least a few (but hopefully many more) good men and women will realize they have been duped.

I know it can be easy to hit the snooze button, roll over, and go back to sleep. But my plea is that beyond a wake-up call, America gets a mighty jolt like the one that swept colonial America nearly three hundred years ago. That can only happen one way. Dylan alludes to this at the end of "When You Gonna Wake Up?" He asks the million-dollar question, "What does God demand?" He then challenges our ideas about God, letting us know our Creator is not a boy Friday who seeks to satisfy us at the snap of our fingers.

But Dylan doesn't just ask the big question, he provides us with the priceless answer. Getting intensely personal, he describes the man on a cross who has been crucified for you—and then tells us simply trusting in His power is all we have to do.

I deeply believe we are at a critical point in history where such a spiritual awakening is needed more than ever, allowing people to step out of the clutches of indoctrination and see the world afresh.

May God richly bless you, your family, and your circle of influence. And may God bless the United States of America.

I appreciate you taking the time to read this work.

Brian Sussman

ACKNOWLEDGMENTS

MY WIFE URGED ME TO write this book. As a Daughter of the Revolution, she is keenly aware that liberty is under attack. She shares my hope that this book will equip readers to take a stand against creeping Marxism and the climate agenda. "Brighty" has been my best friend and biggest fan since we were both eighteen. Having her as a partner makes me feel like I'm the luckiest man in the world.

I also would like to recognize my writing mentor, Johnny Frattarola, for another round of excellent suggestions and edits. You made this book so much better, brother.

Hats off are due to John W. for not pulling any punches with the early drafts of this book, and to Joel C. for keenly dotting i's, crossing t's, and supplying me with encouragement. I also appreciate the scientific and technical input from Dr. Dave, "Jeffty" M., and Tyler Z.

I am indebted to Madeline Sturgeon and her excellent editing team at Post Hill Press, which, for this book, included Holly L. The attention to detail and accuracy during the final phases of copyedits was thoughtful and intelligent.

I am grateful to Greg Johnson at WordServe Literary for his many years of representation. It's an honor to be associated with your agency, Greg, especially given some of the incredible authors you represent.

And finally, a big thanks to Anthony Ziccardi, the Publisher at Post Hill Press, for giving me a wonderful opportunity to make this book available to a significant audience.

ABOUT THE AUTHOR

Photo by Katie Green

BRIAN SUSSMAN IS AN AWARD-WINNING meteorologist, Hall of Fame talk radio host, author, and podcaster. After graduating from the University of Missouri, Brian became a TV meteorologist and science reporter for KPIX Channel 5 in San Francisco, as well as the regular fill-in weather reporter for *CBS This Morning* in New York. Prompted by an urge to speak his mind about politics and current events, he transitioned to talk radio in San Francisco on KSFO-AM, where he hosted *The KSFO Morning Show* for nearly two decades while also filling in for Mark Levin's and Michael Savage's national programs. His first book, *Climategate: A Veteran Meteorologist Exposes the Global Warming Scam*, instantly became a bestseller. He hosts his own podcast called *The Brian Sussman Show*.

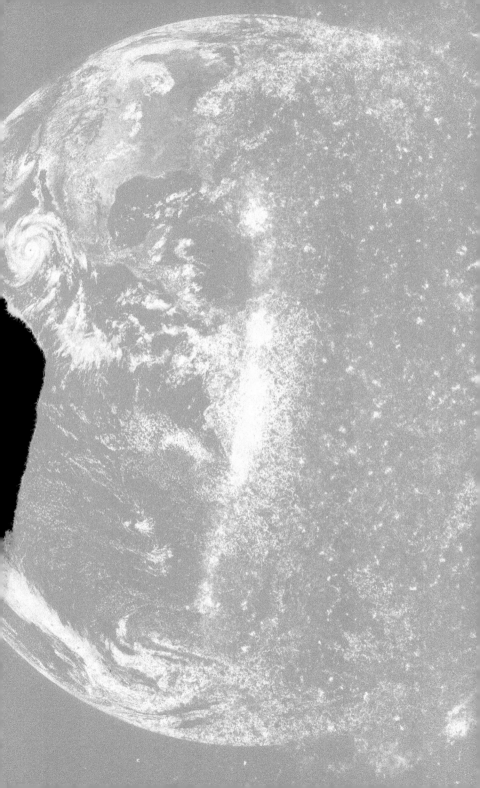